T0258892

Remapping Energopolitics

Emerging concerns and contexts of geological thinking seek to bring out how energopolitical interventions into the geokinetic "unfolding" of the Earth assume *new* dimensions and directions, owing to the complex and evolving intersections between "folds" and "fluxes" of energy in the context of oceans. Written in negotiation with the notion of *energopolitics* articulated by Dominic Boyer, *Remapping Energopolitics* calls for ruling out any epistemic attempt to structure the rhizomatic movements of energy through the transformations of oceans. Aiming to delve deeper into the complex junctures among energy, ocean and earth(*ing*), epistemic ends of Blue Humanities are reworked with the help of geophilosophical reading of some Sri Lankan *minor* writings and in doing so, *Remapping Energopolitics* makes a series of attempts to reconceptualize "energy thinking" in line with the differential and deterritorial grammatology of Deleuzo-Guattarian micropolitics, thereby offering a critique of the structured and stratified understandings of "energy linkages."

Abhisek Ghosal currently works at the Department of Humanities and Social Sciences, Indian Institute of Technology (Indian School of Mines), Dhanbad, Jharkhand. He previously worked as a full-time Assistant Professor at O.P. Jindal Global University (Institute of Eminence), Sonipat, Haryana and at Christ (Deemed to be University), Bannerghatta Road Campus, Bengaluru. He holds an M.A., an M.Phil. and a Ph.D (IIT Kharagpur). His broad areas of research interest include Deleuze and Guattari Studies, Blue Humanities, South Asian Literature, Indic Studies and Energy Humanities, among others. He has published articles in a number of leading academic journals, including *Symploke, New Global Studies, The CEA Critic, Southeast Asian Review of English* and *e-Tropics*.

Routledge Focus on Literature

Writing In-between
Collaborative Meaning Making in Performative Writing
Nandita Dinesh

Billy Lynn's Long Halftime Walk
Flags, Football, and the NFL's "Foxy" Patriotism Problem
Lisa Ferguson

Bosnian Authors in a European Window
A Comparative Study
Keith Doubt

Contemporary Irish Masculinities
Male Homosociality in Sally Rooney's Novels
Angelos Bollas

Creative Writing and the Experiences of Others
Strategies for Outsiders
Nandita Dinesh

Emotionality
Heterosexual Love and Emotional Development in Popular Romance
Arvanitaki Eirini

Digital Culture and the Hermeneutic Tradition
Suspicion, Trust, and Dialogue
Inge van de Ven and Lucie Chateau

Dreams in Chinese Fiction
Spiritism, Aestheticism, and Nationalism
Johannes D. Kaminski

Remapping Energopolitics
Blue Humanities, Geophilosophy and Sri Lankan *Minor* Writings
Abhisek Ghosal

For more information about this series, please visit: www.routledge.com/Routledge-Focus-on-Literature/book-series/RFLT

Remapping Energopolitics

Blue Humanities, Geophilosophy and Sri Lankan *Minor* Writings

Abhisek Ghosal

NEW YORK AND LONDON

First published 2025
by Routledge
605 Third Avenue, New York, NY 10158

and by Routledge
4 Park Square, Milton Park, Abingdon, Oxon, OX14 4RN

Routledge is an imprint of the Taylor & Francis Group, an informa business

ISBN: 978-1-032-62971-1 (hbk)
ISBN: 978-1-032-62973-5 (pbk)
ISBN: 978-1-032-62972-8 (ebk)

DOI: 10.4324/9781032629728

Typeset in Times New Roman
by Deanta Global Publishing Services, Chennai, India

To the fond and loving memory of the late
Prof. Bhaskarjyoti Ghosal, Former Head,
Department of Sanskrit, The University of Burdwan

Contents

Acknowledgement

I take this opportunity to extend my deepest regards to the late Prof. Bhaskarjyoti Ghosal, former Head and Professor, Department of Sanskrit, The University of Burdwan, who played an instrumental role in leading me to take up an ambitious project like this so as to work out an intellectual critique of "energopolitics", taking substantial recourse to Blue Humanities, geophilosophy and "minor" writings.

I wish to thank Prof. Steven Mentz, Professor of English, St. John's University, New York, for kindly writing an engaging Foreword to my monograph. I do believe that Prof. Mentz's profound experience in the field of Blue Humanities expressed through intellectually stimulating reflections in the Foreword makes a precious addition to this work. Scholars and researchers working in the domain of Blue Humanities may find new "lines of flight" to work out for the advancement of Blue Humanities in the future. In short, Prof. Mentz's insightful Foreword both sets the tone of the book at the inception and is quite set to drive intellectual and epistemic investigations of scholars and researchers to myriad openings.

In this regard, I feel immensely indebted to the incessant and selfless guidance of all my teachers who have taught me over the years and still cater to my needs as and when required.

Words actually fall short of describing my beloved teachers' incredible and remarkable contribution to the intellectual developments happening in me. This monograph therefore turns out to be an intellectual upshot of my teachers' full and constant support and leadership. With utmost humility and respect, I simply bow down to their affectual and caring presence that has shaped the kinetic "becomings" of my intellectual development over the time.

I also acknowledge the constant support and encouragement of Arpita Ghosal, my mother, who has been supporting me in every up and down right from the inception of this project. I also admit that during the time of drafting and proofreading, Shankhadeep Ghosal, my brother, provided me with relevant insights so that I could comfortably edit my critical contentions reflected in this work.

I do acknowledge the support and inspiration of my current colleagues at Department of Humanities and Social Sciences, Indian Institute of Technology (Indian School of Mines). I have also got support and help from my former colleagues at O.P. Jindal Global University (Institute of Eminence; Deemed to be University) as well as at Christ (Deemed to be University). Their inspiring words helped overcome occasional exhaustion and weariness associated with the act of writing a full-fledged monograph such as this.

I also want to extend my thanks to the anonymous reviewers and other technical staff for their valuable inputs given at different levels of the preparation of this monograph.

Foreword
The Blue Humanities in a Global View

Steve Mentz

One of the great weaknesses of Blue Humanities as an academic discourse in the early decades of the twenty-first century has been its geographic isolation. The discourse has substantially been driven by scholarship and writing from the Global North. In Anglophone scholarship especially, a great deal of writing in this discourse has been produced from Europe, North America, and English-speaking nations of the Southern Hemisphere including Australia and South Africa. This scholarship, which aims to be global in scope, has noticeably fallen into a pattern of circulation that follows the familiar routes of colonialization, mercantile trade, and global economic exchanges. An important missing region and intellectual community appears in the most populous English-speaking region in the world, the Indian subcontinent. My eagerness to contribute a Foreword to Abhisek Ghosal's *Remapping Energopolitics: Blue Humanities, Geophilosophy, and Sri Lankan Minor Writings* emerges from my own distance from the cultures and waters about which Ghosal writes, as well as how much I have learnt from his distinctive voice.

Where I am writing today, near the shores of the calm but silty waters of Long Island Sound not too far from New York City in the United States of America, the waters of Sri Lanka and the Indian Ocean world appear distant. The northeastern United States has neither coral reefs nor tropical rain forests, and my coastline lacks the ancient global history of long-distance maritime mercantile exchange that has characterized Sri Lanka for more than five millennia.

Reading across that planetary distance, my great pleasure in encountering this book has been discovering the ways in which Ghosal's Blue Humanities appears distinctly *Sri Lankan*. The geophysical environments and ecological systems of the Indian Ocean world distinguish his scholarship from the kinds of Blue Humanities discourses that circulate in other parts of global Anglophone culture. We who think, teach, and write—as well as swim, sail, and surf!—in this critical mode have much to learn from these pages.

In this short Foreword, I will suggest how a view from the Global North and Atlantic waters might productively intermingle with Ghosal's work and help contribute to a globalized and Indian Ocean-informed Blue Humanities.[1]

My opening term, which I adapt from ecological and feminist analyses by scholars such as Stacy Alaimo and Astrida Neimanis, will be "entanglement". My hope in this Foreword will be to entangle my own Blue Humanities with ideas and scholarship from the Indian Ocean world. The concept of entanglement brings together both physical proximity and metaphorical resonance. As Neimanis influentially observes, humans are all "bodies of water", and we share aquacorporeality with other living things, larger bodies such as rivers, lakes, and oceans, and the planetary system as a whole (Neimanis 2017).

The implications of Neimanis's assertion, especially when combined with important concepts such as Alaimo's "trans-corporeality" and insights by Melody Jue, John Durham Peters, and others who treat seawater as media, place the watery blue at the centre of human relationships with our more-than-human environment (Alaimo 2010, Jue 2020, Peters 2015). To recognize our entanglement with water in its multiple forms, from the liquid in our bodies to the vapour in the air, represents an enabling step in Blue Humanities thinking. When we have eyes to see, blue appears everywhere—even if much of the water with which we find ourselves entangled appears less clear flowing blue than muddy brown, biotic green, gleaming white ice, or a rainbow prism of many other colours.

Ghosal's book, like much Blue Humanities scholarship in the 2020s, begins from a basic recognition of water's infiltrating touch. For me as an American scholar, the special force of this work surfaces the varied currents of the blue humanities and the ways that water entangles itself inside and alongside cultural objects. Ghosal's scholarship also speaks to some larger-scale conversations in global water-infused thought that I will outline briefly here.

Global Connections

We know that "oceans connect" on local and global scales, and the human history of the Indian Ocean world is at least five millennia old. The historian Michael Pearson considers the Indian Ocean to be humankind's "oldest sea" because of the antiquity of its trade networks (Pearson 2003).[2] Modern ideas of a global or World Ocean have been central to recent constellations of human interconnectivity, resonantly visualized by the "Blue Marble" photograph of the planet Earth taken by the Apollo 17 crew from space in 1972.[3] Blue humanities scholars have troubled in meaningful ways the burst of optimism that, for some environmentalists in the 1970s, accompanied the Blue Marble image. Elizabeth DeLoughrey's *Allegories of the Anthropocene*, and her continuing argument for a "critical ocean studies", which she distinguishes against the broader "blue humanities", argues for water-thinking that rejects neoliberal economic fantasies (DeLoughrey 2019, 2022).[4] Sidney Dobrin's *Blue Ecocriticism and the Oceanic Imperative* also assumes a global perspective, in particular when he discusses the "protein economy" generated by the fishing industry (Dobrin 2021). The view from the Indian Ocean can productively invert northern-centric visions of globalization, just as attention to the undersea ecologies of shipwrecks and their encrustations, in recent work by scholars such as Sara Rich, Natali Pearson, and Killian Quigley, can reveal new perspectives (Rich 2021, Pearson 2022, Quigley 2023).

These scholars and Ghosal's new book can help reorient understandings of oceans as globalizing vectors.

Black and Blue as Historical Matrices

The cruel history of the North Atlantic slave trade has massively shaped and still distorts the politics, population dynamics, and cultures of the communities that border these waters.

Especially in discussions of the North Atlantic, recent scholarship has expanded ideas about the Black Atlantic into a global vision that places the Black and Indigenous experience of the Middle Passage at the centre of cultural and environmental history. Adding an Indian Ocean perspective to this developing conversation promises to illuminate these vibrant discourses. The slave trade in the North Atlantic remains distinct from colonial expansion in the Indian Ocean world, though some of the same ships, humans, nations, and environmental forces were at work in both. Recent ocean-focused decolonizing work in American Black Studies by scholars such as Christina Sharpe, Dionne Brand, and Alexis Pauline Gumbs (Sharpe 2016, Brand 2001, Gumbs 2020) speaks compellingly to Ghosal's Indian Ocean environment and might enable comparative analyses. The voices that are reorienting oceanic histories in Anglo-Caribbean contexts will make potent interlocutors for Indian Ocean scholarship in ways that can help generate an increasingly global Blue Humanities.

Blue as Capitalist Horror

Many scholars who explore global waters and Anthropocene environments have focused on capitalism's half-millennia of exploitation as key to making sense of oceanic modernity. The collection, *Anthropocene or Capitalocene?* edited by Jason Moore, provides a helpful commingling of key voices in this debate, including the influential ecotheorist Donna Haraway (Moore 2016). Perspectives from the Indian Ocean can usefully supplement Moore's ecoglobalism and other perspectives that have dominated debates about the Anthropocene and many other competing "cenes", from Plantationocene to Capitalocene to Pyrocene, that have been proposed.[5] Historical analyses of capitalism, colonialism, and environmental history, especially those written by scholars in the global North, have a tendency to focus on Northern communities and colonies. The story of waterborne capitalism comes with a different valence from the point of view of the Indian Ocean world and Sri Lanka in particular.

Toward Anthropocene Water

Juxtaposing the competing voices of North and South, East and West, seems necessary to advance the Blue Humanities as a global current of scholarly and artistic communication. That these geographic strands operate to a degree separately may be hard to avoid in practice. But it seems urgent to build connections between different oceanic regions. The larger project of the Blue Humanities ultimately aims to bring into the community all manners of local and global bodies of water. To quote Alice Te Punga Somerville's inspiring words, "I want to ask if Ocean Studies might be better understood if it were itself an ocean: without a singular starting point or origin; endlessly circulating" (Te Punga Somerville 2017). This vision of dynamic flow and non-hierarchical circulation motivates a global Blue Humanities that aims to bring multiple ideas, places, and forms into dialogue. As the Australian-born scholar James L. Smith observes about medieval water cultures, there is a "hydro-social" intersection between water and human cultures that seems important to explore (Smith 2018). A broad vision of Anthropocene water in the twenty-first century and hydro-social forces operating throughout human and nonhuman history can enable new ways of thinking with and alongside watery forms. That dynamic flow characterizes the best possible futures for this fast-moving discourse. Ghosal's Sri Lankan-centred analysis makes an important contribution to this global Blue Humanities. His work also issues an important challenge to scholars and scholarly institutions that reside mostly in Anglo-European West. To know planetary water, we must flow with all its currents.

Ghosal's Energopolitics

The global currents I have been exploring in this brief Foreword find rich expressions in Ghosal's adaptation of Dominic Boyer's concept of energopolitics, which enables a distinctive close analysis of recent Sri Lankan fiction. In attempting what Ghosal calls a "decolonial deconstruction" of the Blue Humanities, this volume clearly does speak to what he calls the "need of the hour". Transforming Boyer's scholarship on spatial transformations and energy flows for a "post-globalization era", Ghosal connects ecological scholarship to waterscapes and literary texts that are not yet well known in the Global North. In connecting eco-materialist ideas to the Vedic suktas, Ghosal integrates distinctive modes of thinking about humans, nature, and energy. Following his imperative that the Blue Humanities must be "interdisciplinary", and following DeLoughrey's insistence that ocean scholars should resist the creeping neoliberalism of the Blue Economy, Ghosal frames a reading of Sri Lankan literature that should have wide appeal in and beyond the Indian Ocean world. His book represents an important contribution to global Blue Humanities scholarship.

Notes

1 The central pages of this Foreword repeat observations I made in a Preface to Claire Hansen and Maxine Newlands's *Critical Approaches to Australian Blue Humanities* (forthcoming Routledge 2024).
2 As I discuss in *An Introduction to the Blue Humanities*, other scholars think of the Pacific, with its ancient migrations, as humanity's oldest sea (Mentz 2024: 65–66).
3 See Mentz 2024: 24 on the Blue Marble image.
4 A developing critique of the Blue Humanities appears in DeLoughrey 2022.
5 A short, and already out-of-date, round-up of roughly two dozen proposed "cenes" appears in Mentz 2019.

Introduction

Why Blue Humanities Matter

In *The World Is Blue: How Our Fate and the Ocean's Are One*, Sylvia A. Earle has cogently reflected:

> Deep coral reefs are being destroyed by new deep trawling technologies aimed at capturing fish that are decades, even centuries, old. The destroyed corals are thousands of years old.   The ocean's pH—the measure of the alkalinity or acidity—is changing owing to increased CO2 that in turn becomes carbonic acid ... So does the ocean matter? Of course the economic uses of the ocean matter—extraction of oil, gas, minerals, fresh water and wildlife, transportation, tourism, real estate enhancement.
>
> (2010, 3)

This striking excerpt clearly points out the fact that the ocean—one of the major support systems of the Earth—stands at risk not only because oceanic resources are inordinately exploited by one section of human beings who hardly care for the complex and densely mediated marine ecosystems which in turn take care of the entire humanity at large but also because the ocean is understood to be a teleological upshot of human practices of social construction which calls for taking into account the ocean as a *constructed reality*, thereby choosing to pay no heed to the ways the ocean is dying at the ground level. Earle's critical reflection can further be worked out in this way that the significance of the material existence of the ocean does not stand confined to catering to the economic needs of human beings living in the world but it *matters* to the ways in which the ocean makes *figural-functional* inroads into the nuanced interactions between human and nonhuman beings conditioned by the appalling trajectories of global warming, ocean acidification, oceanic disaster, (trans)oceanic commercial enterprise, oceanic human migration, global planetary crisis, oceanic nonhuman emergency and so on and so forth.

Interestingly, in *The Social Construction of the Ocean*, Philip Steinberg divulges how the material spatiality of the ocean stands dovetailed into the social spatiality of the environment, which allows voyagers to the oceans, belonging to different spatio-temporalities, to invade, structure and territorialize the spaces of oceanic ecosystem to facilitate cross-oceanic transport

DOI: 10.4324/9781032629728-1

of goods and human capital. Steinberg argues that a social construction-ist approach to the ocean leads one to view the ocean either as a "a space 'outside' society" (2001, 207) or a "friction-free surface" (2001, 207) or "an abstract point on a grid" (2001, 207) or "a repository of fragile" (2001, 207) or "a space constructed" (2001, 207) or "an extension of a land-space" (2001, 207) or "non-possessable but nonetheless a legitimate arena for expressing and contesting social power" (2001, 207–208). He means to say that the oceanic spaces seek to register conflicts and contradictions between social powers, experimentations and exhortations of social spatiality lead-ing to the occurrence of social changes. Thus, the ocean ceases to be a mere topographical referent to a "place" and assumes social spatiality that causes impetus to the ceaseless and rhizomatic movements through social, political, cultural, economic and ecological spaces. Along with viewing the ocean as a *metaphoric-metonymic* referent, a critical reinvention of the ocean under the epistemic rubric of "blue humanities" calls for digging out *new* perspec-tives through which the deterritorial becoming of the ocean that stands *onto-epistemologically* enmeshed with *geokinetic* movements of the Earth can be explored afresh, intending to better respond to the questions of ocean as a *living* geokinetic entity and a *fluid* archival repository so that human engage-ment with the ocean can ethically be remapped along the lines of *differential intensity* or *nomadic-singularity*. In a nutshell, "blue humanities" starts with taking the ocean into account as a *figural-functional* agency and does not end with the ocean itself. It seeks to engage and reengage the ocean with *material-semiotic* dimensions of the socio-political spatiality of ecology spread across the intersection between oceans and coasts, thereby making a critical attempt to *think with the ocean* in order to think through and beyond the ocean. Blue Humanities thus functions as a fluid epistemic *portal* to facilitate one to step into the worlds of intersectional ecology. In fact, Steven Mentz, too, in his recent publication, *An Introduction to Blue Humanities*, has called for an opening up of the epistemic limits of "blue humanities": "Water surrounds us—in our bodies, our neighborhoods, and our planet. The core intellectual challenge of the blue humanities explores how water functions in and across multiple scales. Making sense of disorienting movements across scales and spaces captures the pleasure and ambition of the blue humanities" (2024, xii–xiv). In short, instead of putting critical focus on the varieties of water-bodies, Mentz argues that epistemic delimitation is required to examine the actualities of "bodies of water" to make critical inroads into the multiscaler movements of water across different ecologies.

At this critical juncture, one may stop and think: How do we need to take care of the oceans that selflessly take care of human and nonhuman entities alike? How does oceanic thinking "matter" in one's understanding of social, political, economic, cultural and ecological changes in society? How does oceanic thinking stand mediated through existing critical frameworks that are crucial in figuring out the planetary crisis which intimidates the world at large? Why is it of profound importance to engage oceans in decipher-ing intersectional ecological concerns? Does "blue humanities" seem to get turned out as a Eurocentric epistemic framework or do we need to intervene in conceptual configurations of "blue humanities" from indigenous perspec-tives? What can be done to resist "blue humanities" from being held as a "grand narrative" on histories, politics and cultures of the ocean? Do we need to employ blue humanitarian perspectives to step into the domain of energy to spell out the intensive differentiality of the ocean? Is the ocean a "missing link" between cultures, political events, economic practices and ecological changes? In short, do we need to posit "blue humanities" in the discourses on Critical Ocean Studies to understand nuanced oceanic communication to the anthropocentric views of the ocean? This study seeks to critically consider these questions to contend that the incremental importance of "blue humani-ties" *matters* in comprehending transnational and geo-territorial governance of political economy, cultural plurality and ecological diversity—a combina-torial interplay among these factors is quite important in understanding our deep concerns about planetary emergencies that continuously threaten the world at large. It also aims to argue that a critical reinvention of the ocean in terms of "blue humanities" can offer *new* perspectives both to question human exploitations of endangered marine resources either in terms of "blue trafficking" or in terms of setting up "blue (infra)structural" designs in the middle of the ocean and to sensitise "global" tourists to come to terms with the onerous survivals of coastal people who, to a large extent, depend on the oceanic resources.

Whereas Sylvia A. Earle in *The World Is Blue* speaks of the protection of the ocean to safeguard it from potential planetary disasters: "...how much of the ocean *should* be protected to maintain the vital life-support functions, restore and hold steady of populations of seriously depleted fish and other ocean wildlife, and cope with growing dead zones, ocean acidification, cli-mate change, and massive amount of pollution?" (2010, 187), Elizabeth DeLoughrey in "Toward a Critical Ocean Studies for the Anthropocene" insists that critical engagement with the ocean is the need of the hour to fig-ure out how in contemporary times the uses of the ocean are not limited to mere transports of goods or maritime warfare but it stands inclusive of mili-tarization of oceanic spaces for regulating global economic exercises so as to have political authority over the cultural and ecological practices on the lands: Scholars have called for a "critical ocean studies" for the twenty-first century and have fathomed the oceanic depths in relationship to submarine immersions, multispecies others, feminist and Indigenous epistemologies, wet ontologies, and the acidification of an Anthropocene ocean. This strate-gic military grammar is equally vital for a twenty-first-century critical ocean studies for the Anthropocene because it does not lend itself to an easy poet-ics, the militarization of the seas is overlooked and under-represented in both scholarship and literature emerging from what is increasingly called the blue or oceanic humanities (2019b, 22).

It also indicates how geopolitical militarization of the ocean for meeting certain economic ends gets precipitated in the marine ecosystems, resulting in indelible damage to the complex structures of the marine world. DeLoughrey's perspective on the declining conditions of oceanic life undoubtedly calls for taking governmental measures to provide *ethical care* for the health of the ocean to protect "oceanic habitats" from the potential threats of endangerment. In this regard, one may argue that the multiplicity of the ocean can be understood if one posits *oceanic thinking* in perspectives to explain how the deterritorial movements of the ocean seek to govern the lives of coastal people who live on oceanic resources and determine the political, social and cultural courses of their lives in terms of how the ocean *communicates* with them. In order to explicate the mediatedness of the ocean in *matters* pertaining to postcoloniality, transcorporeality and transreality, one may be reminded of the following excerpts:

> Postcolonial narratives have been instrumental in voicing the gaps, the silences, and often the bitter ironies of Eurocentric renderings of ocean space, precisely by foregrounding those "shadowy, dark bodies" that are more often than not reduced to the status of mere backdrop or obstacle, object or symbol of alterity profiling...
>
> (Bartels et al. 2019, 81)

> We are both of these things, inextricably and at once – made mostly of wet matter, but also aswim in the discursive flocculations of embodiment as an idea. We live at the site of exponential material meaning where embodiment meets water For us humans, the flow and flush of waters sustain our own bodies, but also connect them to other bodies, to other worlds beyond our human selves...the human is always also more-than-human. Our wateriness verifies this, both materially and conceptually.
>
> (Neimanis 2017, 1–2)

These critical reflections suggest how on the one hand postcolonial insights can well be entwined with "oceanic thinking" to examine the cryptic silences of "ocean space" and to decimate stratified approaches to the becomings of the ocean and on the other hand "watery" discourses of embodiment seek to connect supposedly inflexible and impervious "human" bodies with other "bodies of water" enunciated by the ocean, for example. In a nutshell, "oceanic thinking" flattens putatively hierarchized relationships between land and water in general and particularly forges a seamless connection between human and nonhuman agencies circumstanced by differential and intersectional ecological patterns of becoming.

At times, it is argued that "blue humanitarians" tend to speak from Eurocentric viewpoints and thus do not take into account how human beings engage themselves with a series of dialogic interactions with the ocean circumstanced by the prevailing conditions of marginality. In this context, one may be reminded of the following utterances:

> Are we then moving towards a new social history of Indian Ocean networks? A history that looks at not so much about prices and commodities and distribution channels as that of identity formations [but] seems to suggest in the affirmative as historians and anthropologists come together to address notions of a space identity and occupation, of deconstructing discourses on citizenship and entitlement outside the frame of the nation state as well as of the area studies perspective.
>
> (Subramanian 2020, 80–81)

> Indian Ocean is no more a "Forgotten Sea" as it was perceived a few years ago. A new corpus of texts, as Mardewun Adejunmobi puts it, "associated with mainly coastal and island communities sharing in common similar experiences of slavery, indentured labour, colonialism and other deprivations of political and economic rights is part of the new thalassology".
>
> (Sharma 2020, 131–132)

These quotes clearly indicate that scholars working in the field of Indian Ocean Studies resort to new critical perspectives to reinvent the Indian Ocean as a *zone of irreducibility* that seeks to impact the formation of transnational "citizenship" in the context of diasporic dispersal, thereby reconfiguring the discourses of transareality on the one hand and on the other hand they seek to establish that the Indian Ocean ceases to be a "Forgotten Sea" and archives the differential interconnectivity among political, economic and cultural experiences experienced by people in transit. What is important to note here is that in the context of the Indian Ocean, "blue humanitarian" perspectives can be employed to divulge how the Indian Ocean at once turns out to be a "productive site" (Sharma 2022, 133) for rethinking diasporic concerns in the contemporary times and for facilitating the *emergence* of new forms of human-nonhuman relationality. In short, differential "productivity" of the Indian Ocean as a "deterritorial site" for experimentation and exercise gets impacted by the structures, processes and eventualities of 'transareal' patterns of neoliberal governmentality. More importantly, it is true that epistemic borders of "blue humanities" can reasonably be dragged in the context of the Indian Ocean to expound how postcolonial realities in Indian Oceanic zones continue to impact transoceanic movements of people in transit. In other words, Indic inroads into the refashioning of "blue humanities" can tenably be done to widen the epistemic scope of this critical framework. Indigenous knowledge systems in particular need to be resorted to add up g(l)ocal nuances to "blue humanitarian" approaches to the "intensive" transformations of the ocean. In short, decolonial deconstruction of "blue humanities" is the need of the hour to bring up the profound importance of postcolonial indigenous knowledge systems which engage with the discourses of oceanic historicity to lay out how locals dwelling in the coastal regions stand guided and intimidated by the "disjunctive ruptures" in terms of "infrapolitics" (Marche 2012, 3) in the nearby marine ecosystems.

In "Towards a Blue Humanity", Ian Buchanan and Celina Jeffery hold: "we need not only a blue humanities but also a blue humanity—a sense in which we restore our sense of connectedness" (2019, 12), thereby suggesting that epistemic workings and re-workings pertaining to "blue humanities" have to work towards achieving a well-disseminated state of "blue humanity" so that it causes a creative production of a "Plane of Immanence". In continuation with Buchanan and Jeffery, this study aims at stepping into the realm of "energopolitics"—which broadly speaks of spatial politicization of energy flow for meeting certain material ends—inasmuch as energopolitical interventions are nowadays being employed to regulate the *economic-ecological* matrix of the ocean, thereby putting the ocean at risk once again in the time of planetary emergency. Put it in other words, this study is a modest attempt first to get into the domain of energy and then to interrogate the epistemic rigidities of "energopolitics" (2019, 19) as understood by Dominic Boyer in *Energopolitics: An Introduction*.

In the context of South Asia, with the onset of the neoliberal economic framework from the 1990s onwards, followed by the rapid dissemination of technology, exponential growth in the domain of international human migration, steady rise of postnational consciousness among (g)local migrants and transnational trading of "energy capital" embodied by a number of earthly entities and state authorities in association with neoliberal elite traders, the proliferation and dissemination of the global tourism industry have been running the risk of exploiting precious and irretrievable oceanic resources in terms of either allowing g(l)ocal entrepreneurs to build up human infrastructures in the middle of the ocean, intending to extract and exact marine 'energy capital' for transnational exportation to distant locations in the world or beaconing g(l)ocal tourists to create free and unrestrained access to oceanic ecologies which have been at risk. Neoliberal loot of oceanic resources facilitated by the pervasive trajectories of "energopower" leads one to mull over the ruinous and vicious impacts of "energopolitics" on the oceanic resources in particular.

It may be noted that neoliberal exploitation of oceanic resources does not only impact entangled marine environment but also puts the immediate coastal regions under dire ecological threat. Thus, this study seeks to make critical inroads into Dominic Boyer's insular configuration called "energopolitics" by means of employing critical "blue humanitarian" perspectives restructured with the help of the geocriticism of select Sri Lankan "minor" fiction.

In order to carry out the proposed objective, this book is split up into certain chapters to build up epistemological frameworks both to make critical inroads into the nuanced configurations of Boyer's "energopolitics" and Blue Humanities, and subsequently to contextualize critical reflections of select Sri Lankan "minor" fictional narratives. This book comprises five chapters excluding an Introduction and Conclusion. In the opening chapter titled "Nomadic Singularities of the Earth: Negotiating Geophilosophical Reflections", reflective spotlight is put on the epistemological nuances of geophilosophy, intending to examine the interlinkages of geophilosophy with the nuances of "New Earth Politics", Earth sciences, "geokinesis", and "earth thinking" and finally it seeks to arrive at the complex process of "earth(ing)" that stands wedded to the logic of what Deleuze and Guattari call "deterritorialization" in *A Thousand Plateaus*.

The second chapter, "Onto-epistemologies of *Minor* Writing: An Overview", seeks to put emphasis on the manifold figurations of a "minor" writing while particularly laying out transgressivity, micropolitics and differentiality that it seeks to embody. It is argued that a "minor" writing is replete with epistemic interstices which are capable of leading one to figure out its intensive differentiality and are required to tailor the tattered ends of Blue Humanities.

The third chapter, "Cartography of Blue Humanities: Contentions and Contestations", aims to delve deeper into the nuanced and seamless overlapping between the oceanity of the ocean and coastality of the coast thereby remapping the domain of Blue Humanities. Put another way, epistemological strands of Blue Humanities are brought out to underscore its limited scopes of operation, and subsequently, the domain of Blue Humanities is sought to be opened up by means of enfolding coastal ecology, which plays an instrumental role in safeguarding marine ecological resources from g(l)ocal neoliberal loots carried out by the practitioners of "energopolitics". Addition of a coastal ecological framework to the existing grammatology of Blue Humanities is intended to pull the latter from being categorized as an emergent meta-narrative.

The fourth chapter, "Geokinetic Interventions into Matter and Matter(ing): Thresholds of *Energopolitics*", seeks to reflect on the *allagmatics* of energy through geological, political, and cultural frameworks of the Earth. In doing so, it takes Dominic Boyer's "energopolitics" (2019, 19) to task for failing to put up counter-discursive frameworks to the neoliberal loots of oceanic resources. In short, trans(formative) and trans(formational) potentials of "matter" are taken into account to divulge how Boyer's "energopolitics" (2019, 19) ends up becoming a State politics inasmuch as it stands grounded in Foucauldian biopolitics, thereby offering a critique of the former.

The fifth chapter, "Sri Lankan *Minor* Fiction: Earth(ing), Energy Flows and Oceanic Ecologies", seeks to offer a detailed contextualization of the critical arguments pertaining to the intersecting trajectories among energy, ocean and earth(ing) as reflected in select Sri Lankan "minor" fiction—Romesh Gunesekera's *Reef,* Roma Tearne's *Mosquito,* Shyam Selvadurai's *Swimming in the Monsoon Sea* and Chandani Lokuge's *Turtle Nest.*

In the Conclusion, the domain of Energy Humanities is succinctly explored to contend a couple of arguments: The notion of "energy thinking" could be worked out to preclude energy being subjected to strata and territories on the one hand and on the other hand, *figural-functional* dynamics of energy need to be critically taken up to widen human engagement with the fluxes and folds of energy.

I Nomadic Singularities of the Earth

Negotiating Geophilosophical Reflections

A new cry resounds: the Earth, the territory and the Earth!
(*A Thousand Plateaus: Capitalism and Schizophrenia*,
Deleuze and Guattari 1987, 338)

The Earth is undergoing a period of intense techno-scientific transformations.
If no remedy is found, the ecological disequilibrium this has generated will
ultimately threaten the continuation of life on the planet's surface.
(*The Three Ecologies*, Guattari 2000, 27)

the human as a geological agent, whose history could not be recounted from
within purely humanocentric views. This agency was not autonomous and
conscious, as it was in Thompson's or Guha's social histories, but that of an
impersonal and unconscious geophysical force, the consequence of collec-
tive human activity The globe, I argue, is a humanocentric construction; the
planet, or the Earth system, decenters the human.
(*The Climate of Histories*, Chakrabarty 2021, 3–4)

In the post-globalization era, when climate change is still looming large on the living and nonliving entities on the Earth, human beings remain engaged in a series of anthropocentric experimentations to exert human power over non-human beings, hoping that it is by means of subjugating nonhuman beings, human beings shall be able to put a check on the declining conditions of climate change which happens to be one of formidable geological challenges that humanity faces today. Following Chakraborty's timely intervention into the consequences of climate change, it can be argued that human beings, one of the most powerful and insensitive geological agents on the Earth, cannot but try to think with the Earth whose deep histories stand intricately enmeshed with the much shorter human histories on the Earth so that human efforts to exploit so called abundant resources of the Earth can be checked for wellbeing of the Earth in general and particularly for the comfortable and smooth lives of other nonhuman entities.

DOI: 10.4324/9781032629728-2

Therefore, a critical reengagement with the becomings of the Earth needs to be taken up to build up a geokinetic epistemology backed up by geophilosophical reflections of both Western and Eastern origins to facilitate individuals to better respond to the threats of climate change manifested through various kinds of geological disasters including oceanic catastrophe.

Considering the immense importance of Earth both as a geological referent and a deterritorial force in deciphering the nuanced manifestations of climate change, this chapter is framed to examine different epistemological interstices of geophilosophy while critically responding to the geophilosophical question: How does the Earth think it is? In a nutshell, the objective of this chapter is chiefly twofold: epistemological interlinkages of geophilosophy with the nuances of "New Earth Politics", Earth sciences, "geokinesis", and "earth thinking" are first intended to be laid out and in doing so, it is by working out the complex process of "earth(ing)" charged with the quanta of deterritorialization, a geophilosophical epistemology is intended to be constructed to help individuals understand how human beings need to reengage with the transformative and transformational movements of the Earth so as to hold out against the threats of climate change.

Geophilosophy: A Critical Overview

> *All philosophy is geophilosophy.*
>
> ("Geophilosophy as the End of Philosophy",
> Colebrooke 2022, 169)

. . . the geocritical emphasis on space, place, and mapping correlates strongly to the conviction among spatially oriented critics that space is of the utmost social importance. ("Introduction: Ecocritical Geographies and Geocritical Ecologies and the Spaces of Modernity" Tally and Battista 2016, 2)

To put it in simple words, geophilosophical thinking can be understood as a philosophically- driven way of looking into the nuanced becomings of the Earth, which finds material reflections through the ceaseless transformations of the different components of the Earth. As Eileen Crist rightly put forward in *Abundant Earth*: "Connections between Earth's ecological formations are as seamless as the life-forms of its sundry biomes are diverse" (2019, 214), thereby underscoring the profound importance of the immanent materiality of the Earth, which lays down an open ground for the seamless interactions and intersections among different earthly entities. Taking recourse to Crist's intervention, it can be contended that in order to grasp densely mediated and interconnected operativity of the Earth, one cannot but reckon with geophilosophical insights which are capable of spelling out "trans(in)fusive" potentials and deterritorial becomings of the Earth. So far as geophilosophical thinking is concerned, one cannot but refer to *What Is Philosophy?* where Gilles Deleuze and Felix Guattari have densely formulated the operative logic of "geophilosophy" in the following terms: "[geophilosophical] thinking takes place in the relationship of territory and the earth" (1994, 85). It suggests that geophilosophical thinking stands in the tension between the "territory" and "the Earth" and thus a geophilosopher has to take stock of how territory and the Earth stand as "two components with two zones of indiscernibility—deterritorialization (from territory to the earth) and reterritorialization (from earth to territory)" (1994, 86). It means that a geophilosopher finds the deterritorial movements of the Earth in the tension between "deterritorialization" and "reterritorialization". This is also implicative of the fact that geophilosophical thinking caters instrumental support to an individual who is interested in exploring the two mutually exclusive processes—"deterritorialization" and "reterritorialization"—something which holds secrets of the movements of the Earth. Geophilosophy actually helps one differentiate territory from the Earth—whereas the first being "the emergence of matters of expression" (Deleuze and Guattari 1987, 315), the second is the material embodiment of the deterritorialized. It is also true that the territory and the Earth share a complex bonding between them in the sense that whereas territory refers to *"a reorganization of functions and a regrouping of forces"* (1987, 320) and bears "coefficients of deterritorialization" (1987, 326), the Earth is "the intense point at the deepest level of the territory" (1987, 338) and therefore, the "earth is certainly not the same thing as the territory" (1987, 338). In short, whereas Deleuzo-Guattarian understanding of "territory" is actually a discursive embodiment of how the process of reterritorialization gets actualized, Deleuzo-Guattarian understanding of "the Earth" is in fact a geophilosophical incarnation of deterritorial flows that drive the Earth away from being caught up by the 'forces' of territorialization.

In connection with the salient tenets of geophilosophy laid down by Deleuze and Guattari in *What Is Philosophy?* one may be reminded of "gaia philosophy"—an inclusive philosophical standpoint which is capable of divulging layered inactions and intersections among different earthly bodies and their consequential impacts on the geological becomings of the Earth, and therefore, it pertains to the conceptual strands of geophilosophy. The word "gaia" means "Mother Earth" in Greek and refers to bio-geo-eco-spheres. Gaia, the Greek goddess of the Earth, seeks to put the constitutive elements of the Earth namely water, soil and air, among others, and living organisms in order thereby explaining the homeostatic system of it.[1] James Lovelock[2] happens to be an esteemed expert in the field of "gaia philosophy", who sums it up as a philosophical instrument for exploring the Earth as a self-sustaining, self-regulating and self-referential system. It suggests that different components of the Earth work together in maintaining the functionality and materiality of the Earth as a composite system. Gaia philosophy helps one unravel how the Earth works as a "machinic assemblage" in the sense that the growth and development of several living organisms on the Earth stand in tandem with the geokinetic movements of the Earth and consequently, the Earth grows up as a composite whole. To put it in Deleuzo–Guattarian fashion, just as in a machine, different component parts seamlessly function together thereby producing a "force" for a machinic activity, the Earth performs a "desiring-machine"[3] as it were in the sense that the kinesis of the Earth results in the "production" of "intensive" and "extensive" becomings of the Earth itself in general and particularly, entails physical transformations of all the earthly entities.

At this point, one may take a pause and mull over a couple of questions to get into the nuances of geophilosophical thinking: How does "gaia philosophy" work at praxis? Does "gaia philosophy" actually help one interrogate the following assumption that the Earth is a "static" system? These questions could be responded to by drawing a reference to "Gaia and Philosophy" where Dorion Sagan and Lynn Margulis hold that "gaia philosophy" provides a "scientific view of life on Earth" and thus, it helps one comprehend the "earthly factual". "Gaia philosophy" is actually premised on "the circular logic of life" on the Earth and takes "a radical departure" from the Darwinian view that on the Earth, life is conditioned by and adapts to a static environment (1997, 145). Instead, "gaia philosophy" can lead one to insist that the "temperature and composition" of the Earth are in fact "regulated" by "biota" (Sagan and Margulis 1997, 145). It suggests that the Earth stands as a synergetic system in which different constituting "organs" seamlessly function together to turn it into a "body without organs" or an embodiment of deterritorialization. In short, "gaia philosophy" lays down a *scientific foundation* for geophilosophical thinking to rule the roost in the domain of Earth Studies.

In order to reinvent the literary underpinnings of geophilosophical thinking, one may tenably draw the notion of "geocriticism" that stands grounded in the subtle interplay among—"spatiotemporality", "transgressivity" and "referentiality" (Westphal 2011, 122). In the "Preface" to *Geocriticism: Real and Fictional Spaces*, Robert T. Tally Jr. explains that geocriticism primarily helps one understand both the way "literature interacts with the world" and that "all ways of dealing with the world are somewhat literary" (2011, x). In a way, geocriticism stands as a critical way of exposing the interface between the world and literature by means of taking recourse to *multifocalization* and *polysensoriality* (Tally Jr 2011, x). Geocriticism thus takes into account ontology of a territory, transformation of a territory and how a territory moves along the lines of "space, place and literature" (Tally Jr 2011, xi). In other words, geocriticism is a geo-centred way of looking into the representations of spatiality in literary narratives and thus is of profound importance. Here, one may argue that geocriticism can reasonably be subsumed as the *literary logic* of geophilosophy in the sense that it makes a passage ready for geophilosophy to slip into the "minor" becomings of literary narratives. Therefore, suffice it to say, both "gaia philosophy" and "geocriticism" lay down *scientific and literary* grounds for geophilosophical thinking thereby empowering the latter to facilitate individuals to better address human-induced changes to the Earth and its ruinous corollaries.

Indic Interventions into Geophilosophical Thinking: Negotiating Vedic Śuktas

The concept of geophilosophical thinking is generally understood to be a Western epistemological framework, owing to the great contributions of Martin Heidegger, Gilles Deleuze, Felix Guattari, Ben Woodard, Thomas Nail, Reza Nagarestani, Martin Hägglund, Gary Genosko, Claire Colebrook, Mark Bonta, John Protevi and so on. Opposed to this
general perception, it can be contended that one may find hitherto untraded tradition of geophilosophical reflection in Vedic śuktas which need to be studied to look into the nuances of Indic critical inroads into the epistemological formulation of geophilosophy. In Vedic śuktas, seers did not use the word geophilosophy per se but epistemological constituents of geophilosophical thinking can well be mapped in Vedic śuktas. Whereas Western scholars primarily engage themselves in the movements of the Earth to find out geophilosophical dimensions of it, Vedic seers started to examine the extensive and intensive movements of the Earth much before geophilosophical thinking came into being in the Western epistemologies. For instance, in Vedic times, when science was not as advanced as it is today, Vedic seers formulated the concept of *dyāvāpṛthivī* to examine how the becomings of the Earth stand densely entwined with the material activities of human beings. In different Vedic śuktas, Vedic seers attempted to lay down conceptual dimensions of *dyāvāpṛthivī* to claim that a critical understanding of *dyāvāpṛthivī* is of profound importance to spell out material-metaphorical engagement of human beings with the Earth. *Dyāvāpṛthivī* consist of two words—*dyaus* and *pṛthivī*—a combination of which speaks of how the Earth stands inclusive of *dyuloka*, meaning bright place thereby referring to the abode of gods and goddesses, and *bhuloka*, meaning this physical world where human and nonhuman entities live together. It means that the breadth and width is not limited to where human and nonhuman beings reside. Rather, the expanse of the Earth, the views of Vedic seers, is limitless. Unlike Western geophilosophical models, Vedic seers resorted to sacred hymns to narrativize the figural-functional dimensions of *dyāvāpṛthivī* and at times deified it to make people think with the Earth. Here, one may be reminded of some Vedic śuktas to demonstrate how *dyāvāpṛthivī* was comprehended and extolled in effusive terms:

> I PRAISE with sacrifices mighty Heaven and Earth at festivals, the wise, the Strengtheners of Law.
> Who, having Gods for progeny, conjoined with Gods, through wonder-working wisdom bring forth choicest boons
>
> (Rgveda Griffith trans. 1.159.1)

> Extolled in song, O Heaven and Earth, bestow on us, ye mighty Pair, great glory and high lordly sway,

Whereby we may extend ourselves ever over the folk; and send us strength
that shall deserve the praise of men...

(Rgveda Griffith trans. 1.160.5)

Wide, vast, and manifold, whose bounds are distant,-these, reverent, I
address at this our worship,
The blessed Pair, victorious, all-sustaining. Protect us, Heaven and Earth,
from fearful danger.

(Rgveda Griffith trans. 1.185.7)

These quoted excerpts suggest that *dyāvāpṛthivī* is made up of both Heaven
and Earth—an epistemic conglomerate that accounts for how Vedic seers
took up a holistic approach to make critical inroads into the immanence of the
Earth. *Dyāvāpṛthivī* is deified not only to make Vedic common people believe
in the enormous power of it in resolving different kinds of anthropocentric
issues and concerns but also to underscore how it safeguards human and non-
human entities alike from potential ecological disasters and dangers. It is clear
that *dyāvāpṛthivī* spans across the physical Earth and more importantly, it
constantly corresponds to *dyuloka* to keep the becomings of the Earth locked
up in a sort of homeostasis. One may also cogently put forward that deifica-
tion of *dyāvāpṛthivī*, an Indic addition to the discourses of geophilosophy, was
intended to serve certain purposes: First, it could effectively deter Vedic common people from exploiting abundant resources of the Earth out of their sheer
fear of the Almighty; second, it could constantly encourage Vedic common
people to make ethical engagements with the Earth so that *dyāvāpṛthivī* did
not get offended; third, it would inculcate in them a practice of thinking with
the Earth, which is nowadays missing in anthropocentric ways of knowing the
Earth; fourth, it speaks of how one may rely on discourses of spirituality to
discuss the transformations of the Earth.

In addition to it, one may also be reminded of other geophilosophical
reflections that pertain to the sustainable becomings of the Earth:

May Heaven and Earth, the Mighty Pair, bedew for us our sacrifice,
And feed us full with nourishments.

(*Rgveda* Griffith trans.1.22.13)

Thornless be thou, O Earth wide before us for a dwelling-place Vouchsafe
us shelter broad and sure.

(*Rgveda* Griffith trans.1.22.15)

O Heaven and Earth, with accord promoting, with high protection as of
Queens, our welfare, Far-reaching, universal, holy, guard us. May we, car-
borne, through song be victors ever.

(*Rgveda* Griffith trans. 4.56.4)

To both of you, O Heaven and Earth, we bring our lofty song of praise,
Pure Pure Ones! to glorify you both.

(*Rgveda* Griffith trans. 4.56.5)

These Indic geophilosophical reflections attest to the fact that Indic conceptualization of *dyāvāpṛthivī* is of manifold entity that accounts for the safe,
healthy and sustainable lives of humans and nonhuman beings on Earth.
Interestingly, *dyāvāpṛthivī* is understood to be a guiding force that is at once
protective in nature and at times allows human beings to freely engage with its
ceaseless becomings expressed through the physical transformations of Earth.

At this point, one may stop and think: how does *dyāvāpṛthivī* make a valuable addition to the discourse of geophilosophy? What is the relevance of
dyāvāpṛthivī when modern physics advances in conformity with different
branches of science? Epistemic formulation of *dyāvāpṛthivī* does not stand
restricted in the mere conceptualization of its figural aspects; rather, it underlines how human beings need to develop ethically sound connections with
the deterritorial movements of the Earth so that human thinking can onto-
epistemologically be tied up with the thinking of the Earth that finds reflections in the sharp tension between deterritorialization and reterritorialization.
It means that epistemization of *dyāvāpṛthivī* calls for ethically sound actions
on the part of human beings who indiscriminately pounce upon seemingly
abundant yet limited resources of the Earth, thereby making room for severe
climate change to take shape. In addition, a practice of Vedic geophilosophical thought pushes one to take into account the actions taking place in the
realm of "outside". This means that the notion of *dyāvāpṛthivī* seeks to impact
geopolitical activities carried out by human beings in that human geopolitical
interventions need to stand in tandem with the geophilosophical unfolding
of the Earth; otherwise, it may result in the eruption of a series of ecological catastrophe. As *dyāvāpṛthivī* is held responsible for keeping the Earth
green and productive, it calls for forging a thinking tradition with the geokinetic unfolding of the Earth. Whereas Deleuzo-Guattarian geophilosophical
insights seek to explain the "geokinesis" of the Earth in terms of the dialectical
interplay between "territory" and "the Earth", Vedic geophilosophical reflection were well ahead of its time and showed the way the Earth corresponds to
the world of "outside" in terms of energy flows and constantly makes a series
of "intensive" self-adjustments in commensurate with changes happening in
the external terrain. In "Reconfiguring Asian Modernity: Negotiating Tantric
Epistemological Traditions", Abhisek Ghosal and Bhaskarjyoti Ghosal
cogently argue: "Tantric sādhakas usually engage in exploring the cryptic
potentials of Śakti, which pervades all living and nonliving entities on earth"
(2022, 37). What this implies is that In Indic Knowledge System (IKS), there
is a tradition of knowing the becomings of the Earth in terms of tapping into
its inherent rhizomatic energy flows. Whereas Vedic seers began to engage the
concept of *dyāvāpṛthivī* to lay out "energy linkages" prevalent in geokinetic

transformations of the Earth, *Tantric* sādhakas later on kept up this tradition by means of practising "tantric paths" for enlightenment. Thus, it is by employing *dyāvāpṛthivī* in reality, one may be able to figure out how the Earth thinks in terms of deterritorial energy flows which find material reflections in being rendered as veritable "body without organs". Plus, a pragmatic understanding of *dyāvāpṛthivī* tells how Vedic seers called hierarchical interplay between Heaven and the Earth into question; and quite ingeniously, they put both of them on the "Plane of Immanence", thereby pointing out the fact that the Earth itself does not corroborate hierarchized anthropocentric thoughts.

In *Rgvedic Dyāvāpṛthivī and Modern Science: A Relevant Study*, Shrimanta Chattopadhayay holds:

> The origin of sky-watching system of the modern science can well be traced back to old Vedic literature and the modern scientists can traverse the universe of truth that is enunciated in the Vedas. In fact, the profundity of thoughts of the seers is pervaded by and surcharged with the poetically mystic imagination.
>
> (9)

This clearly indicates how the modern sky-watching system can at times refer back to Vedic configuration of *dyāvāpṛthivī* which calls for taking into account *dyuloka* and *bhuloka* together. Besides, much before modern physicists recognized the importance of "quantum entanglement" in figuring out the movements of the Earth, Vedic seers could understand the importance of bringing nuanced and conjoined workings of Heaven and Earth together, that stand embodied in the geophilosophical framework called *dyāvāpṛthivī*. In fact, in "Vedic Thought and Modern Science", Madhuchanda Kaushik pertinently observes:

> The Vedic method of acquiring knowledge uses both the objective as well as the subjective techniques and hence knows the total knowledge which satisfies the criteria of being science Matter is not a primary entity; it is a product of space-time continuum. Vedic seers see the universe as an inseparable web, whose interconnections are dynamic and not static. Modern physics also believes that the universe as such is a web of relations and like Indian mysticism, it has recognized that this web is intrinsically dynamic. The dynamic aspects of matter arises in quantum theory as a consequences of the wave-nature of subatomic particles and is even more essential in relatively theory, as we shall see, where the unification of space and time implies that the being of matter cannot be separated from its activity.
>
> (2015, 140–141)

Kaushik is quite right in claiming that as the Vedic understanding of matter underlines onto- epistemological fluidity of it, Vedic scientific queries stand grounded in both objective and subjective understandings of the external reality—a critical standpoint that pertains to the quantum entanglement of "space-time" continuum. In short, Vedic geophilosophical inroads into the geokinetic movements of the Earth can, to an extent, be upheld by a modern physical standpoint called "quantum theory".[4]

"New Earth Politics": Making Geophilosophy Work

At this point, one may take a stop and mull over a couple of questions: What exactly is "New Earth Politics"? How can geophilosophical strands facilitate one to grasp the nuances of "New Earth Politics" and its corollaries? In order to expound "New Earth Politics", one may straightaway refer to "Living on a New Earth" where Simon Nicholson and Sikina Jinnah have emphatically argued that we, the humanity, have entered a "New Earth"—"an Earth 2.0 on which the human signature is everywhere and in desperate need of humane and insightful guidance" (2016, 1). What they mean to say is that New Earth refers to a new dimension of the Earth, which is marked out and regulated by human activities. New Earth is basically grounded in the following two premises—human beings have (re)emerged with power to alter "the functions of earth systems" and global environmentalism has taken an unprecedented "scale and speed" in tandem with human advancements in the field of techno-communication (Nicholson and Jinnah 2016, 3). It is argued that "New Earth Politics" starts to be unfolded with the rolling out of economic globalization and global technological advancement. It is true that New Earth gets shaped by "technological innovation" and "economic wealth" (Nicholson and Jinnah 2016, 6) and most importantly, is concerned with the dire and dreadful consequences of both "technological innovation" and "economic wealth" on the developments of the Earth itself. In a nutshell, "New Earth Politics" is critical of the exploitative human interventions into earthly resources, subsequently leading to the damage and depletion of earthly entities, and at times lays down radical ways of putting checks on the oppressive human activities and bringing the Earth back in its natural synergetic system of operation. In "The Changing Shape of Global Environmental Politics", Ken Conca reflects that "the politics of a New Earth … (includes) the ongoing restructuring of the world economy … (and) the continuing decline of what might loosely be termed the world's 'environmental middle class'" (2016, 23). Conca means to say that the dynamics of "New Earth Politics" is subject to global economical, political, technological and ecological alterations. Whereas economic globalization entails "the construction of fundamentally global platforms of production and the parallel manufacturing of global consumer aspirations" (2016, 24), and technological advancements result in the augmentation of developed communicational algorithms, "New Earth Politics" seeks to actualize *deterritorial reforms* in the political governance of the ecology of the world.

Here, one may tenably contend that geophilosophical insights can, in many ways, facilitate one to delve deeper into the fluid configuration of "New Earth Politics". For instance, geophilosophy helps one understand how earthly movements of planetary entities straddle between territory and the Earth. It means that social, political, geological, economical and cultural movements on the Earth are stuck in the *intermezzo* of two "zones of indiscernibility". Geophilosophical interventions into "New Earth Politics" lead one to grasp how human beings engage themselves in territorializing the Earth into "strata" and the Earth in turn tends to transgress human codifications and stratifications. Besides, it also helps in establishing *symbiotic* connections between living and non-living entities on the Earth—an important dimension of "New Earth Politics". Geophilosophy, apart from exposing the ontology of "New Earth Politics", comes in handy in explaining how the essence of "New Earth Politics" lies entrenched in the tangles of geology, politics, economics, ecology, topography and spatiality. Unlike the following opinionated assumption that the Earth is a static and inflexible phenomenon, geophilosophy lays bare the inherent dynamicity and unconventionality of "New Earth Politics", which keeps on rolling with the passage of time. "New Earth Politics" calls for a radical critique of crass anthropocentric overtures on different components of the Earth and lays down fluid and modifiable patterns of ecological governance suited to the Earth's natural tendency to get over human geopolitical codings. In short, "New Earth Politics" gets fuelled by geophilosophy which, together with other geo-centric approaches, prepares readers to come to terms with the nuanced functioning of the Earth as a "multiple". Put in other words, geophilosophy can divulge how the "multitude" can group together to speak up against the atrocities of neoliberal elites against the Earth. Earth has a healing mechanism of its own and it requires a good amount of support from the "multitude" which can make use of globally "networked political activities" (Parr 2018, 199) to *de-stratify, de-hierarchize* and *de-code* the Earth.

Geophilosophical Interventions into the Earth Sciences: Critical Reflections

Geophilosophysimultaneously informs Earth sciences how the kinesis of the Earth stands deeply entrenched in the logic of deterritorialization and at times opens up the new possibilities for Earth sciences by laying emphasis on the overlapping trajectories of geo-mathematics, geo- physics and geometry. Geophilosophical inroads into Earth sciences entail how the Earth gets ungrounded time and again by its "tensional vegetality"[5] (Nail 2021, 135) on the one hand and by the continuous interactions between living and non-living entities on it, on the other hand. In *On an Ungrounded Earth: Towards a New Geophilosophy*, Ben Woodard pertinently reflects that the act of "ungrounding" the Earth is intricately associated with "decay" and "process" in that the act of decay puts the Earth in a perpetual process of "ungrounding" which results in the making and remaking of the Earth's interiority and exteriority simultaneously. In a way, the Earth proceeds through "the logic of succession". Woodard opines that it is because of its "limitropic porosity", the Earth simultaneously performs "nature's product and also its productivity" (2013, 7–8). In short, the Earth slips into a process of *(en)foldment* the more it embraces "enveloping-developing" (*The Fold*, Deleuze 1993, 8) dynamics. Interestingly, *(en)foldment* turns the Earth into an *ensemble* of problems which are unstable and erratic. To put the *(en)foldment* of the Earth in terms of "smooth infinitesimal analysis model", one may put forward that "intensive" and "extensive" movement of the Earth is "differential" in nature and is conspicuously subject to the differential interactions between folding and unfolding. The more the Earth gets folded up, the more the phenomenon of unfolding gets increased simultaneously, resulting in the tangential *(en)foldment* of the Earth. Put in other words, it is because of the interplay between "intrinsic singularity" and "extrinsic singularity" (Duffy 2013, 16) that the Earth gets *(en)folded* and thus, the "differential" progression of the Earth entails "an essential singularity" (Duffy 2013, 30) whose "lines of rupture" actualize the "minor" becomings of the Earth as it stands as an assemblage of *function, derivation* and *force* or *flow, fold* and *field*, as it has been elaborately articulated by Thomas Nail in *Theory of the Object*.

Strands of geophilosophy are quite instrumental in explaining geo-mathematics as a set of mathematical algorithms which are needed to explain how the "differential movements" of the Earth take place in general and particularly how structural changes pertaining to the interiority and exteriority of the Earth occur. In short, following geophilosophical insights, it can be contended that geo-mathematics is in fact involved in rationalizing the "organization of the complexity of the system Earth" (Freeden 2015, 7) and thus can help one to figure out how the Earth "functions" as a complex and singular system conditioned by the "interacting physical, biological, and chemical processes transforming and transporting energy, material, and information" (Freeden 2015, 6). This contention can further be worked out in the way that geo-mathematical thinking, circumstanced by the strands of geophilosophy, leads one to discern the *co-extensive interconnectedness* among different constituting elements of the Earth.

Tenets of geophilosophy also stand wedded to the critical concerns of geophysics which seek to deal with the complex movements of the Earth's physical properties like lithosphere, asthenosphere, mesosphere, atmosphere, biosphere, hydrosphere, and so on, conditioned by human activities and offer scientific methods of looking into the becomings of the Earth. In *The Mantle of the Earth: Genealogies of a Geographical Metaphor*, Veronica Della Dora ingeniously argues that the mantle of the Earth is marked by "mutability and unpredictability" (4) and hence is quite interesting. Any movement of the mantle results in the *intensive* and *extensive* becoming of the Earth. Geophysical intervention into the Earth's differential intensity in particular can help one bring out how different physical properties of the Earth pass through a series of transformations while interacting between themselves—a phenomenon that attests to the proccessual aspects of the Earth. It also helps one comprehend how human-induced disruptions break down natural mechanisms of the Earth system thereby putting the Earth at risk.

Geophysical thinking soaked in geophilosophy leads one to construct a kind of discourse that seeks to rail against human shaping and reshaping of the Earth by means of either establishing illicit mining industries in the oceans or planting machineries on the Earth to extract geo-resources for monetary gain. It also helps one unpack the inherent system of the Earth, which stands threatened by exploitative human activities. Thus, in *The Unconstructable Earth: An Ecology of Separation*, Frederic Neyrat rightly rejects "geo-constructivist" perceptions to uphold the "neo-organicist" assumptions on the Earth. The Earth is a large "living being" (2019, 166) which has a self-governing and self-sustaining mechanism that conditions the growth and development of other organisms as well. Thus, it is clear that epistemic overlapping between geophysical thinking and geophilosophical insights is, in fact capable of spelling out the dialectical interplay between strata and flows embodied by the "supple segmentarity"[6] of the Earth.

Doctrines of geophilosophy are densely interconnected with the paradigms of geometrical thinking in many ways. Following Michel Serres's emphatic claim in *Geometry: The Third Book of Foundations*, "What is geometry, again and finally? The measurement of the earth" (2017, 163), it can be argued that geometry has close affinity with the measurement of the Earth in the sense that it provides different mathematical insights to take stock of the becoming of the Earth. Whereas geo-mathematics and geo-physics advocate for an observation-measurement-orientated approach, geophilosophical inroads into the patterns of geometrical thinking lay down different measurement techniques so as to help one map the *enfoldment* of the Earth and thus no one can have "the slightest idea or perception of an earth without geometry" (Serres 2017, 211). Epistemic crossovers between geophilosophy and geometry can be worked out in such a way that a combination of both geophilosophical and geometrical insights not only explains the material transformation of the Earth but also pushes one to delve deeper into the "abstract and intelligible understandings of individual spaces and relationships between spatial objects" (Rawes 2008, 3). Rawes's understanding of geometric thinking, to some extent, pertains to Deleuze and Guattari's formulation of "diagrammatic" progression in that, like a diagram, geometric progression of earthly entities cannot be territorialized inasmuch as they always tend to slip into the passage of "deterritorialization".

Geokinesis of the Earth: Interfacing *Geo-containment,* *Geo-history,* and *Limitrophy*

> . . . *we must envisage . . . a new pact to sign with the world: the natural con-*
> *tract. (The Natural Contract* Serres 1995, 15)

At this critical juncture, one may stop and think: What does "geokine-sis" of the Earth entail? How do epistemic crossovers among "geo-contain-ment", "geo-history" and "limitrophy" pertain to the "geokinesis" of the Earth? In *Theory of the Earth,* Thomas Nail appositely puts forward that the Earth "is much more like a process than it is like a stable object" (2021, 19). Following Nail's critical perspective, it can be argued that the Earth is con-tinually trans(form)ing in compliance with the changes of its constitutive ele-ments. This argument can further be worked out by referring to what Nail calls "geokinetics"—the kinetic theory of the Earth that spells out the dis-tinctive (form)ation and form(ability) of the Earth. "Geokinetics" is in fact premised on three interconnected analytics—"the flow of the matter", "the fold of elements" and "the circulation of planetary fields" (Nail 2021, 20). Taking recourse to Nail's "geokinetics", one may tenably posit that the Earth in actuality "flows" and thus denies any sort of human coding and stratifica-tion. The "flow" of the Earth is upheld by the dynamic matters of the Earth. For instance, soil, an important component of the Earth, is made up of the Earth and (re)makes up the Earth too by "regulating the atmosphere, support-ing life, and storing water" (Nail 2021, 24). It is because of the fact that the matter of the Earth "flows, folds and circulates" (Nail 2021, 25), the Earth denies permanent spatio-temporal settlement.

The interplay between strata and flows embodied by the geokinetic move-ments of the Earth may be explained by considering what Ben Woodard calls "geo-containment" which refers to the "vitalism of the inorganic" life that is quite "generative" and "more temporally stable" than organic life on Earth (Woodard 2013, 64). To put it in other words, "geo-containment" refers to a set of strategies that are employed to explore how the Earth is time and again *weaponize*d to serve human purposes. For example, human beings often resort to "burial" as a ritualistic method to dispose of human remains, but this leaves a mark on the surface of the Earth and sheds off human traces as it passes through structural reforms over time. "Geo-containment" thus turns out to be a human plan to territorialize the Earth but it fails to resist the pat-terns of "geokinesis" which, by means of overpowering and overruling human plans, keep the Earth *flowing, folding* and *circulating* through time and space. Jan Zalasiewicz rightly contended in *The Earth after Us: What Legacy Will Humans Leave in the Rocks?*:

> The surface of the Earth is no place to preserve deep history The surface of
> the future Earth will not have preserved evidence of contemporary human
> activity ... For the Earth is active. It is not just an inert mass of rock [nor]

will it be inert. It is a dynamic system, powered from inside by the heat generated from the radioactivity within its interior.

(2008, 14)

Zalasiewicz basically tries to point out the "geokinetic" aspects of the Earth, which erase human traces off the surface, rendering alterations in the geological configuration of the Earth. In a nutshell, "geokinesis" of the Earth finds vivid reflections in terms of its intensive *dynamic systematicity* which makes it supple, malleable, fluid and processual in nature.

Interestingly, the notion of "geokinesis" stands connected to the epistemic underpinnings of "geohistory". In *What Is Philosophy?* Deleuze and Guattari have in fact brought up the concept of "geohistory" while spelling out the operation of geophilosophy. They reflect: "Philosophy is a geophilosophy in precisely the same way that history is a geohistory from Brandel's point of view geography wrests history from the cult of necessity in order to stress the irreducibility of contingency" (1994, 95–96). "Geohistory" thus connotes that history is necessarily entrenched in geography which, in turn, provides "lines of becoming" for history. Put in other words, history has to go along with geography simply because both are mutually exclusive to each other. Thus, geohistorical thinking seems to be useful for understanding how the Earth *flows, folds* and *circulates*. Geohistorical thinking provides an epistemological model which helps one lay bare the fact that the historical progression of the Earth stands in tune with its geographical restructuring either caused by inherent tension between energy and entropy or by any external human force. It is quite clear that the history of the Earth cannot be figured out without taking recourse to its geographical patterns. Geohistorical thinking also puts emphasis on the fact that the Earth stands as a "plane of consistency" where history resides in the geography and vice versa. Geohistorical thinking can empower a geophilosophical approach to the Earth in that it directs one to take note of how physical properties of the Earth undergo temporal and tensional shifts to give an impetus to the "geokinetic" movement of the Earth. "Geohistory" thus lays down *a materialist foundation* which equips one to investigate "minor" becomings of the Earth through spatio-temporalities.

In addition to the establishment of geo-historical connections with geophilosophical thinking, it is argued here that the epistemic strands of the "geokinesis" of the Earth can further be reinvented by taking into account what Jacques Derrida calls "limitrophy" in *The Animal That Therefore I Am*:

> *Limitrophy* ... [not just only refers to] what sprouts or grows at the limit, around the limit, by maintaining the limit, but also what *feeds the limit*, generates it, raises it, and complicates it ... [it works] certainly not in effacing the limit, but in multiplying its figures, in complicating, thickening, delinearizing, folding, and dividing the line precisely by making it increase and multiply.

(2008, 29)

Derrida means to say that "limitrophy" is a critical way of getting engaged with the "limit" not to question its polemical position but to "multiply" it for bringing out heterogeneous discontinuities between organic and inorganic substances on the Earth. What is interesting about this critical formulation is that it upholds the principle of multiplicity and seeks to delve deeper into the living and non-living substances thereby exposing how "the limit" gets tendered and complicated when it passes through "thickening" and "folding". Epistemological study of the "limit" is of profound importance in understanding geophilosophical dimensions of the Earth in general, and particularly in decoding how compositional elements ceaselessly interact with different fluid categorical limits. Derridean intervention into the "limit" reminds one of Deleuze's understanding of stratification which is essentially a human construct manufactured to put the Earth into territories. *Limitrophic* insights can help one figure out the crux of the "geokinesis" of the Earth in that it makes one aware of the (form)ation and form(ability) of "the limit"—ceaseless making and unmaking of which regulates "minor" becomings of the Earth. Derridean formulation, that is, "limitrophy" encourages one to augment and multiply the figuration and (trans)figuration of "the limit" so as to make out heterogeneous formations and functions of it, whereas Deleuze and Guattari pertinently point out that a discourse of "multiplicity" always guides one to delve deeper into the layers of signification inculcated in an entity: "the laws of combination therefore increase in number as the multiplicity grows" (1987, 8). Clubbing the discourses of "multiplicity" and "limitrophy", it can be argued that a "limit" is a multiplicity, which unfurls itself as it gets folded up. In the context of "geokinesis", a "limit" works as a "multiplicity", and it starts off to get complicated as the Earth slips into a series of "minor" transformations, leading the "geokinesis" of the Earth to be "compositional, hybrid, and additive" (Nail 2021, 62) in nature.

Earth Thinking and Earth(ing): How do Nomadic Singularities of the Earth Respond to the Phenomenon of Climate Change?

> the Earth—the Deterritorialized. This body without organs is permeated by unformed, unstable matters, by flows in all directions, by free intensities or nomadic singularities.
>
> (*A Thousand Plateaus*, Deleuze and Guattari 1987, 40)

In order to delve deeper into the epistemic facets of "geokinesis" further, the notion of "earth thinking" needs to be conceptualized by drawing a reference to what Sam Mickey understands as "whole Earth thinking" which happens to be "a way of understanding and responding to the challenges of planetary coexistence" (2016, 2). "Earth thinking" therefore can be comprehended as a discursive strategy to "account for interdependent and self-organizing dynamics of Earth's systems, including its living systems (biosphere) and its systems of water (hydrosphere), rock (lithosphere), air (atmosphere), and even human consciousness (noosphere)" (2016, 5). "Earth thinking" can tenably be employed to expound seamless interactions between different earthly materialities—be it human or non-human entity, thereby interrogating the ineffectuality of "human" stratification and territorialization. As Mickey rightly put forward: "No matter how different we are, whether human or nonhuman, we are earthlings. We share a common dwelling, a singular home. There is only one Earth" (2016, 5). In short, "earth thinking" stands soaked in what Ranjan Ghosh calls "earth-consciousness and earth-directedness" in "Globing the Earth: The New Eco-logics of Nature" (2012, 8). Here, one may argue that "earth thinking" actually orients one to figure out the intensive processuality of the earth, expressed in the forms of "folding", "flowing" and "circulating". This contention can be backed up by bringing in the concept of "body without organs" (BwO) which, as Deleuze and Guattari have formulated in *A Thousand Plateaus*, connotes a kind of body that is in fact not stripped of organs but that does away with "organism" and "organization" (1987, 30). Unlike a normative body which consists of heterogeneous organs working together in an organized fashion to make the body function properly, a BwO is "singular" in that it prescribes a full eradication of systematic arrangement of functions done by organs and a total embrace of "asignifying particles" or "pure intensities" (1987, 4). A BwO is premised on "multiplicities" which in actuality hold the becomings of the Earth firmly. Deleuze and Guattari argue that a multiplicity does not have any "subject" or "object", but has only "determination", "magnitude" and "dimension" (1987, 8).

Considering the passage of "pure intensities" of the Earth through the process of *(en)foldment* and "geokinesis", one may tenably argue that the Earth could be taken into consideration as a BwO—a pure multiplicity. The process of Earth(ing) thus entails the flows of "unstable matters" and "free intensities" (1987, 40). Deleuze and Guattari rightly reflect: "the Earth ... is a body without organs" (1987, 40) thereby implying that the Earth, instead of getting struck by "stratification", "coding" and "territorialization", tends to slip into "de-stratification", "decoding" and "deterritorialization". This leads one to arrive at that geophilosophical insights that actually help one to bring out how the Earth, instead of being invested with "codes" and "strata", seeks to undergo a self-sustaining process to traverse temporalities. "Earth thinking" can therefore be held as an investigative strategy to arrive at the complex process of earth(ing) that accounts for the "co-intensive" and "co-extensive" "lines of becoming" embodied by the "geokinesis" of the Earth itself. In a nutshell, the "earth thinking" is quite able to actualize what Bernd Herzogenrath called "white ecological" thinking which is "a way to think the environment as a negotiation of dynamic arrangements of cultural and natural forces, of both nonhuman and human stressors and tensors, both of which are informed and 'intelligent'" (2013, 1), thereby facilitating one to grasp the deterritorial transformation of the Earth or the process of earth(ing) that stands wedded to what Deleuze and Guattari call a "plane of consistency" steered by the forces of "deterritorialization".

At this critical juncture, one may stop and mull over the following questions: How does the Earth find geokinetic momentum while indulging in deterritorial becoming? How do exploitative human actions impact bio-geo-eco-semiotic interactions between the Earth and climate change? Where do nomadic singularities lead the Earth? These questions are important to take up not just only to bring out how geophilosophical insights can be employed to withstand the strikes of climate change but also to foreground how human beings need to revise and redesign their manifold connections with the Earth so as to stave off the impact of climate change at bay. In *Singularity: Politics and Poetics*, Samuel Weber argues: "what is 'singular' can only be conceived or experienced as an event that quite literally comes out (ex-venire) of that which it follows and exceeds. But it [singularity] is a repetition that is composed not just of similarity, but of irreducible difference" (2021, 2). Weber is of the viewpoint that the notion of singularity cannot be reduced to a mere process of repetition although singularity emerges out of it inasmuch as singularity consists of "irreducible difference" that makes it appear different from what is called "individual". Following Weber's contention, it can be argued that rhizomatic or nomadic flows of singularity can account for the Earth's deterritorial becomings at the level of intensity that get exteriorized in the domain of the "outside" in terms of physical alterations. This means that the Earth slips into a state of geokinesis conditioned by its inherent nomadic singularities that pull it back from being caught up by the agents of territorialization.

It is because of flows of nomadic singularities, that the Earth engages itself in the process of self-adjustment to come to terms with the physical alterations happening to it due to the onset of climate change. As climate change is understood to be the result of the unwarranted exploitations of natural resources done by a section of human beings who hardly care for the way the Earth selflessly stands protective of its inhabitants and irresponsibly attempt to exert power to control the deterritorial movements of the Earth, it is high time that human beings redesign their ethical engagements with the Earth not only to put a check on the declining conditions of climate change but also to rely on the the enormous power of the Earth to get rid of this present predicament—a critical standpoint that Vedic people ingeniously picked adopted to let the Earth take care of its inhabitants democratically. To put it in other words, instead of exerting sheer power on the Earth, human beings need to develop a habit of thinking with the geophilosophical underpinnings of the Earth so that human beings can refrain themselves from further exacerbating the overarching phenomenon called climate change. In order to execute this geophilosophical strategy to take on climate change, human beings can put "plasti(e)cological thinking"[7] to practices so that an understanding of how the Earth stands superior to the cumulative efforts of human beings grows.

Notes

1 Toby Tyrrell is one important thinker who has referred to Kirchner's "The Gaia Hypotheses" while spelling out variants of "gaia". Tyrrell contends that "homeo-static Gaia" can be understood as a system consisting of the stabilizing effects of biota (2013, 6).

2 Lovelock expatiates on the enormous importance of "gaia" in understanding how the Earth system stands as a "singular" and "self-regulating" system comprising human and non-human components. Lovelock argues in "Reflections on Gaia": "Gaia is a new way of organizing the facts about life on Earth, not just a hypothesis waiting to be tested" (2004, 3).

3 Following the observation on "desiring-machine" by Deleuze and Guattari in *Anti Oedipus: Capitalism and Schizophrenia*, one may argue that it is "a break in the flow in relation to the machine to which it is connected, but at the same time is also a flow itself, or the production of a flow, in relation to the machine connected to it" (1983, 38).

4 In "Parallels in Physics and Philosophy", Susmita Bhattacharya shows how mod-ern physics stands connected to geophilosophical understanding of one's existence on the Earth: "Our scriptures declare – 'Yathā Piṇḍey Tathā Brahmāṇḍey'. The microcosm contains the whole macrocosm. The microcosm preserves in it all the knowledge and entire potentialities of the macrocosm. The world is homogeneous, and modern science shows beyond doubt that each atom is composed of the same material as the whole universe" (2015, 39).

5 In *Theory of the Earth*, Thomas Nail has actually expatiated "tensional vegetality" in the following terms: "Vegetal communication is kinetic, haptic, performative, and dynamically netted in a vast mineral and atmospheric network of intersecting material flows. Vegetality is not reducible to a single object or body but is the whole tangled and in-tensional web of fluctuating signals and responses" (2021, 150).

6 In *A Thousand Plateaus*, Deleuze and Guattari hold: "it is a *line of molecular or supple segmentation* the segments of which are like quanta of deterritorialization" (1987, 196). It suggests that "supple segmentarity" works by the logic of deterrito-rialization and therefore can defy and deny categorical territorialization.

7 In *Plasti(e)cological Thinking: Working out an (Infra)structural Geoerotics*, Abhisek Ghosal puts forward how "plasti(e)cological thinking" can be worked out in the context of climate change that looms large with the passage of time. Ghosal holds:

> An understanding of "plasti(e)cological thinking" can lead the humanity to incorporate the massive role of non-human agency in transforming the Earth. "Plasti(e)cological" insights can help ecocritics to address how plastics, among other toxic and terrible pollutants, can transform the geohistory of the Earth. I shall also open up how the sphere of ecology can be stretched off to non-human agencies. Stagnancy of ecology can be removed by means of consider-ing "plasti(e)cological thinking" which underlines the "supple segmentarity" of ecology.
>
> (2023, 26)

This excerpt shows the importance of taking the transhistorical fluidity of the Earth into account in redesigning plans to take on the challenges posed by climate change. In short, human beings need to progressively withdraw their sheer domi-nance on the Earth by letting other earthly species be with the becomings of the Earth. Humanity needs to rest on the nomadic singularities of the Earth which stand capable of taking both human and nonhuman agents out of the potential dangers caused by climate change.

II Onto-epistemologies of *Minor* Writing

An Overview

Introduction: "(Minor)ity" Is an Incomplete Project

Being = Difference = the New.
("Deleuze and the Production of the New", Smith 2008, 151)

"minor" language is the instrument par excellence *of that destratification.*
("Foreword: The Kafka Effect", Bensmaia 1986, xvi)

Nowadays, scholars working in the field of Critical Humanities engage discourses of "minor" to contest how "minor" is "normatively" viewed in the critical parlances. Deleuze and Guattari happen to be two important thinkers who understood the importance of refashioning the epistemic configuration of "minor", aiming to call conventional understanding of "minor" into question.

Interestingly, the notion of "minor" cuts across social, political, cultural, economical and other aspects of the society in that "minor" gets governed by its inherent micropolitics which helps it avoid the "forces" of territorialization. Contestation of usual understanding of "minor" is intentionally done to open up how the discourses on "minor" take into account the factor of "processual unfolding" that shall later on be taken up in the subsequent paragraphs. In short, enormous potentials of "minor" could be reinvented by taking a dip into the onto-epistemological heterogeneities that stand guided by a series of micropolitical intensities. Thus, this chapter aims at spelling out the nuances of a "minor" writing by means of examining the overlapping trajectories of transgressivity, micropolitics and differentiality.

Simply speaking, "minority" is, to a large extent, configured as a definitive category which typifies either a group of people belonging to any race, class, creed, community and ethnicity and among others, who are less in number and inferior in power, or a bunch of entities which are no longer important. In other words, political configuration of "minority" entails grouping together of a limited number of people or entities, who/which cannot vie with a majority of people or entities. It interestingly suggests that the discourse of "minority" is understood either in terms of positing it *against* the "majoritarian" counterpart or engaging it in a dialogic interaction *with* the discourse of "majority".

DOI: 10.4324/9781032629728-3

"Minority" is thus understood as a *quantifiable category* which helps one figure out "non-majoritarian" perspectives of a limited number of people in a given context. Besides, conventionally speaking, a number of pejorative epithets like "inferiority", "infirmity", "powerlessness", "marginality", negligibility", "insignificance" and so on are frequently attached to the configuration of "minority" so as to turn it into a *derogatory metaphor*. Sometimes, in different socio-political and cultural contexts, the discourses of "minority" are subtly configured in such a way that it can help a ruling authority to size up dissenting people dwelling in a marginal territory under this category to take punitive actions against them. It points to the fact that state authority often resorts to "majority-minority dynamics" to politicize the dissents and differences of marginal people and thus uses "minority" as a *categorical vector* to run down any counter-strategic movement. Thus, a certain configuration of "minority" facilitates a certain political governance to rule the roost.

Structuring "minority" in compliance with the political governance of a territory results in the disintegration of "minority" into several epistemic ensembles. For example, "minority" ceases to be a political tool to quell the rebellion of marginalized people when it is employed to question the veracity and verifiability of majoritarian discourses. It means that a ruling authority cannot have full control over the manipulation of "minority" in serving several purposes.

Discourses of "minority" could resistively bounce back at the ruling authority when marginalized people take recourse to their so-called "minority" position to take on governmental buffets.

Moreover, "minority" is often deliberately understood as an apolitical configuration which has nothing to do with the running of a state. Contrary to this, political uprisings under the banner of "minority" often prove quite unsettling for a government which offhandedly mistakes them as "inferior" and "insignificant". It suggests that the configuration of "(minor)ity" is not confined in the socio-political domain and cannot be subjected to categorical framework. For instance, philosophical inquiry into the epistemic configuration of "(minor)ity" suggests that it has been in a state of "incompleteness" and has been continuously folding and unfolding itself in tandem with the contemporary socio-political alterations. In other words, the epistemic configuration of "(minor)ity" is bound to be an "incomplete" project in the sense that there are layers of significance embedded in it and more importantly, "(minor)ity" in actuality seeks to transgress human codifications and stands as an effective multiplicity.[1]

Ontology of *Minor*: What is a *Minor* Writing?

> *A* minor *architecture is political because it is mobilized from below, from substrata that may not even register in the sanctioned operations of the profession.*
>
> (*Toward A Minor Architecture*, Stoner 2012, 4)

At this critical juncture, one may stop and mull over a couple of perturbing questions: On epistemological grounds, is "minority" the same as "minor"? Does there exist any ontological difference between "minoritarian" and "minoritarianism"? Whether Stoner's understanding of "minor" corresponds to the common perception of what "minor" is but one can certainly think of taking recourse to the distinct ontology of "minor" as explicated by Gilles Deleuze and Felix Guattari in *Kafka: Toward a Minor Literature* to get a nuanced understanding of "minor".

Deleuze and Guattari reflectively argue that "minor" is certainly incongruous with the configuration of "minority" as a definitive political category. Unlike the latter, epistemic configuration of the former done by both Deleuze and Guattari is fluid and dynamic in nature.

The word "minor" does not have epistemic proximity and congruity to "minority" in the sense that "minor" is neither a derivational upshot of "minority" nor a referent to a category of linguistic framework. Plus, whereas "minoritarian" is an individual who upholds the ideological principles of "minoritarianism", "minoritarianism" hints at an ideological position as opposed to that of majoritarianism. Deleuze and Guattari have put down a list of three significant aspects of "(minor)ity" while spelling out the characteristics of a "minor" literature: "...[the] three characteristics of "minor" literature are the deterritorialization of language, the connection of the individual to a political immediacy, and the collective assemblage of enunciation" (1986, 18).

According to Deleuze and Guattari, one may find a sort of "deterritorialization of language" in a "minor" literature in the sense that "minor" literature does not give in "linguistic territoriality" (1986, 25). It means that "minor" literature takes escape from the usual mode of language use and allows a writer to tinker with unconventional language uses. "Minor" literature does not exhibit the formal use of a language and gives room to linguistic experimentations thereby encouraging a writer to practice "deterritorialization" linguistically. It connotes that a "minor" literature allows a writer not to follow the grammatology of any language and to take up fluid linguistic paths to avoid the snares of territorialization.

Due to the cramped space in a "minor" literature, a writer is veraciously forced to connect individuals to the political context. It means that in a "minor" literature, literary figures have to negotiate their immediate political contexts time and again to move forward in the fictional space. In other words, everything in a "minor" literature is political in the sense that literary figures take recourse to political measures either to sort out their immediate problems or find out potential openings to get rid of their individual troubles. "Minor" literature is undoubtedly one of the finest sites of political reconstruction, for it allows both a writer to politicize the literary and the literary figures to take part in politics. Distinct politicality of a "minor" literature is conducive to the manifestation of radicality in terms of thought and action—a state to which the writer and the literary figures yield. It is important here to note that unlike other literary works, "minor" literature is radically political for being transgressive, processual, unfinalizable and machinic in nature, and in a nutshell, "minor" literature is in a state of becoming. Moreover, radical politicality of a "minor" literature does not let it be held in terms of structured and institutionalized epistemologies inasmuch as a "minor" literature is perpetually in a process of becoming.

Within a "minor" literature, there is no possibility for "an individuated enunciation" (1986, 17), for each individual concern is bound to end up in a collective assemblage of enunciation. It happens because a "minor literature" is machinic in nature and comprises a number of intersections and interfaces between different individual concerns. In other words, in a "minor" literature, individual concerns can neither be singled out of the pervasive political backdrop nor can be differentiated on the ground of individuation. In other words, a "minor" literature tends to deny structural "organicity" and "organization", and to celebrate the huge potentials of assemblage and immanence. Unlike an organization, an assemblage calls for deterritorialization which seeks to condition radical political becoming of a "minor" literature. At this point, one may stop and think: Is a "minor" literature synonymous with a marginal literature or a popular literature? Deleuze and Guattari hold that "minor" does not, in any way, designate "specific literatures" but refers to the "revolutionary conditions" for every literature within the stream of great literature. It means that "minor" is not a definitive category but a set of "revolutionary conditions" which help a literature stand aloof from being subjected to institutionalized norms.

A "minor" literature thus can be considered as a "collective multiplicity" (1986, 18) which vouches for the enunciation of collective assemblages. Intensive machinic setup of a "minor" literature at once allows it to evade the snares of territorialization and human overcodings and at times leads it to embrace differentiality, processuality and transgressivity. In a sense, any major literature has to pass through "minor" to be revelatory and revolutionary in nature. What is interesting about "minor" is that it consists of a number of ensembles which lead a literary work to contest *categorical mapping*—a sort of definitive exercise which entails codification of a piece of work in terms of language, narration and individuation. In this context, one may be reminded of Ken Gale's thought-provoking article "Writing Minor Literature: Working With Flows, Intensities and the Welcome of the Unknown" where it is tenably argued that "an engagement with the entanglements and the more complex differentiation of intensities" (2016, 306) help a "minor" literature not to yield to fixities and rigidities. Thus, an epistemic investigation of the ontology of "minor" renders it a veritable body without organs (BwO) which functions as a "collective enunciation".

Transgressional *Minor* Sciences: Microfluidics, *Mathegraphy* and Nomadology

> *Process thinking is a way to think about that which is otherwise possible, and in so doing capture the ongoing emergent nature of organizational life. It is a philosophy of possibility and creativity, without ignoring the restraints of the existing.*
>
> (*A Process Theory of Organization*, Hernes 2014, 66)

Following the observation by Deleuze and Guattari in *A Thousand Plateaus*, one can plausibly argue that "minor science" is "continually enriching major science, communicating its intuitions to it, its way of proceeding, its itinerancy, its sense of and taste for matter, singularity, variation, intuitionist geometry and the numbering number" (1987, 485). They meant to say that "minor" science has the potential to define major science in a new way. But one may stop here and mull over a couple of questions: What is "minor" science? How does it operate? "Minor" science can be understood as a "subsystem or an outsystem" (1987, 105) which, unlike "a constant and homogenous system" (1987, 105), is characterized by a potential becoming. Put it in other words, "minor" sciences celebrate singularity and arecontingent upon mathematical thinking in the sense that a broken line tends to get turned into a curve which always seeks to evade any sort of geometrical territorialization. Whereas major sciences are marked out by definite formation and form(ability), "minor" sciences slip into geometrical interstices, intending to step into the process of "becoming-minor" (1987, 104). In other words, "minor" science is replete with slippages, cuts, folds, lines and ruptures.

Microfluidics is generally understood as a complex scientific mechanism through which flows and fluid jets are precisely manipulated to perform certain activities. Generally, a fluid jet is constrained into a geometric space but precision-guided channelization of fluids helps one understand how laminarity works at praxis. In *Basic Theory and Selected Applications in Macro- and Micro-Fluidics*, Clement Kleinstreuer attempts to understand microfluidics as "a study of the transport processes in microchannels" (2010, 351), Bastian E. Rapp is another important thinker who holds in *Microfluidics: Modeling, Mechanics and Mathematics* that microfluidic flows are essentially "laminar" (2017, 3) in nature, and "capillarity" (2017, xxix) happens to be one of the important corollaries of any microfluidic transmission. Opposed to these critical standpoints, one may argue that microfluidics, in some ways, celebrates the layering down of water flow into strata which do not interact with themselves. Whereas microfluidics is restricted within the stratified flows of fluids in a strictly closed system, "minor" goes beyond the limits of systematic stratification and calls for a potential becoming in terms of geometric interstices and stands capable of redefining fluidics in a new way. Besides, whereas microfluidics speaks for dialectical relationality between laminarities, "minor" pushes one to think of post-dialectical relationality which seeks to account for potential becoming. Unlike microfluidics, which bears conceptual congruities with fluidics, one of the important branches in the field of Physics, "minor" seeks to turn away from the codifications of major sciences and to lay down a new epistemological framework to spell out the deterritorialized standing of "becoming". Moreover, whereas microfluidics is involved in "multiplexing" layers of water flow, "minor" science is capable of "de-multiplexing" the flows of different kinds thereby letting each one assume a sort of singularity.

Deleuze and Guattari have interestingly argued in *A Thousand Plateaus* that although "minor" science is contingent upon mathematical thinking, it speaks of "mathegraphy" more than "mathelogy" (1987, 364). It means that "minor" science is more grounded in geometrical interstices which help one map out a territory in terms of mathematical reasoning than in hardcore mathematics which is primarily associated with quantification. "Mathegraphy" stands as a "projective" and "descriptive" geometrical method of drawing a "diagram" which seeks to generate "lines of ruptures" instead of getting caught in a rigid and frigid structure. "Mathegraphical" thinking is made up of flows and fluidities, which are (un)programmatic in nature, and it stands divested of fixities and rigidities. Another interesting feature of "mathegraphical" thinking is that it seeks to de-stratify a territory and "to open up a new function" (1987, 134) thereby. "Mathegraphy" actually " operates by *matter*, not by substance; by *function*, not by form" (1987, 141), hence "diagrammatic" (1987, 141) in nature. "Mathegraphy" actually informs "minor" science in many ways and stands as one of important mathematical foundations of it. For example, it provides "diagrammatic" models of thinking to "minor" science so as to help it evade the process of stratification and engage in functions and fluidities. It also stands as an embodiment of geometrical deterritorialization that works as an instrument in sustaining the transgressivity of "minor" science.

Nomadology, an important offshoot of "minor" science, is premised on the philosophy of *war machine* which, being an exteriority of State Apparatus, bears "spatiogeographic", "arithmetic" and affective" aspects. It mainly deals with how nomads traverse from points to points and stand as potential threats to the State, for they do not belong to State-sponsored political categorisations. Movement of a nomad is bound to be transgressive and dismissive of all codes and territories; the existence of a nomad can thus be traced in an *intermezzo*—a (trans)positional territory which makes it easier for a nomad to stand out. A nomad usually traverses an uncharted path and refuses to give in any sort of programmatic coding. Nomadology can inform "minor" science in many ways. For instance, it lays down the model of *war machine* which is required to call the codifying operations of major sciences into question; to prove that "flows and currents", fluidity and flux help "minor" sciences compete with the functioning of major ones. As the exteriority of a "war machine" lies in its own "metamorphoses", it cannot be contained by a State-authority.

Similarly, one may plausibly argue that "minor" sciences in actuality work as a *war machine* to defy and deny the overarching dominance of major sciences. Working of "minor" sciences is nomadic in nature, for it seeks to exceed the periphery of duality. "Minor" science is overtly critical of the typical "organization and organism", for these put limits on the free movements of it. It usually functions on "a plane of consistency" forged by nomadological principles, and takes up "asignifying" steps to make "diagrammatic" progression. Unlike major sciences which basically tend to "follow" several patterns, movements and configurations, "minor" sciences seek to "reproduce" something new out of the existing participants. Exteriority of "minor" science is *coextensive* with the becoming of the Earth and thus is of profound import in figuring out the processuality of the Earth.

Figurations of *Minor*: *Quantum Flow*, *Micropolitics* and Differen*t*iation

"Minor" takes up a wide range of manifestations on "the Plane of Consistency" and its rhizomatic figurations can be located in the state of "trans-"—(trans) fusion, (trans)location, (trans)politicality, (trans)formation, (trans)duction, (trans)mission, (trans)(in)fusion, (trans)ference, (trans)gression, (trans)mutation, (trans)cription, (trans)action, (trans)genesis, (trans)oceanic, (trans)versal, and so on. It suggests that heterogeneous figurations of "minor" are not confined in the strata and codes; rather, stand in conformity with the lines of becoming and hence revolutionary in nature. Nuanced and entwined figurations of "minor" can be figured out in terms of segmentarity, micropolitics and differen*t*iation—a fluid combination of which plays an instrumental role in the actualization of "minor" at praxis.

In *A Thousand Plateaus*, Deleuze and Guattari emphatically argue that human beings are spatially and socially "segmented" animals and the figures of segmentarity are either "linear" or "circular" or "binary". It has two types—"primitive segmentarity" and "supple segmentarity"—whereas the former accounts for the continual formation of different inflexible strata, the latter helps one spell out the plasticity of different material forms. It is undoubtedly true that the figuration of "minor" is not divested of the lines of segmentarity simply because these shape up the revolutionary becomings of a "minor". At this point, one may note that it is the "supple segmentarity" that is instrumental for explaining why it is almost next to impossible to put "minor" into strata. Being accompanied by the coefficient of deterritorialization, "minor" seeks to transgress mechanisms of stratification and to indulge in a sort of "quantum flow"—which can be figured out as the "decoded flow" of a sign or a degree of deterritorialization (1987, 219).

Following the principle of "quantum flow" as laid out by Deleuze and Guattari in *A Thousand Plateaus*, it can be put forward that both the political radicality and the radical politicality of "minor" are embedded in its uninterrupted indulgence in a "quantum flow". This contention can further be worked out in this way that supple segmentarity and "quantum flow" work hand in hand in refraining "minor" from being subjected to the politics of Majoritarianism.

Here, one may be reminded of Ranjan Ghosh's formulation, "dynamical plasticity"[2] which can help one comprehend why it is so difficult to pin down "minor" into strata. In *Trans(in)fusion*, Ghosh observes that "dynamical plasticity" connotes a sense of possession and settlement and a pleasant dispossession and unsettlement simultaneously—"agentiality and pleasant dispossession" (2021, 2). In other words, the essence of "dynamical plasticity" lies in its multiple "trans-belongings" (Ghosh 2021, 3)—an entangled state where this "essence" is always in "possession in transit" mode. This is very true in the case of "minor" in that the being of minor is always in "possession in transit" mode and thus is able to surpass the snares of stratification.

Put it in other words, it is precisely the "dynamical plasticity" of "minor" that does not let it get held up in a territory and conditions the continual unfolding of it along the "lines of becoming". Keeping Ghosh's observation in mind, it can be argued that "minor" can be subsumed as a *trans- force field* as it were in the sense that on the one hand, it can help one figure out *molecular geometry* of "minor" and on the other hand, it helps one expose revolutionary and revelatory potentials of "minor".

In *A Thousand Plateaus*, Deleuze and Guattari irrefutably argue that everything is "political" in nature and every politics is " simultaneously a *macropolitics* and a *micropolitics*" (1987, 213). Whereas "macropolitics" is characterized by extensity and exteriority, "micropolitics" is steeped in intensity and interiority. Unlike "macropolitics", "micropolitics" seeks to uncover how an entity subtly takes resort to *differential intensity* to evade the limits of territorialization and to account for the endless unfolding of it. "Micropolitics" usually functions at the level of intensity and stands as a logic of deterritorialization. In a way, it is the "micropolitics" that helps an entity from being designated as an instance of rigid segmentarity and leads it to find out micropolitical ways of "becoming-minor". Whereas "macropolitics" pushes an object to comply with the principles of rigid segmentarity, "micropolitics" liberates an object from the territorial overcodings and leads it to get driven by its own transgressivity and dynamicity. Besides it, "micropolitics" also speaks of *molecular* alterations which are responsible for the "trans-individuation" of an object. Thus, it is of profound import in understanding how a "minor" text, being driven by its inherent *differential intensities*, slips into a series of micropolitical adjustments and readjustments to evade the macropolitics of territorialization. Radical politicality of a "minor" text can be figured out by the help of "micropolitics" in that as "micropolitics" is governed not by "the smallness of its elements but [by] the nature of its "mass"—the quantum flow" (Deleuze and Guattari 1987, 217), it lays down effective interstices to examine the nuances of radical politicality embedded in a "minor" text.

It is rather more interesting that "micropolitics" cannot be understood in isolation and stands complexly weaved together with "macropolitics". Deleuze and Guattari cogently argue that "it is evident that the segmented line (macropolitics) is immersed in and prolonged by quantum flows (micropolitics) that continually reshuffle and stir up its segments" (1987, 218). This loaded statement precisely spells out how "micropolitics" is solely responsible for carrying *intensive becoming* of a "minor" text—which ultimately triggers changes in the *molar configuration* of that "minor" text. "Micropolitics" is held profoundly important for elucidating why every becoming is "a block of coexistence" (Deleuze and Guattari 1987, 292). The notion of "minor" does not prioritize future over the past or vice versa; rather, actually celebrates *co-extensive* and *co-intensive* presences of both "lines of becoming" and "becoming-revolutionary" conditioned by the trajectories of "micropolitics".

Differentiality happens to be one of the hallmarks of a "minor" text in that it both facilitates a "minor" text to sidestep strata and codes, and drives it to slip into its own dynamicity and transgressivity. Differentiation is the very algorithm by means of which a "minor" text gives in the processual unfolding of itself. Differential progression of a "minor" text gets reflected in its continual intensive becoming. It is thus by means of the logic of differentiality, a "minor" text is able to show resistance to *categorical mapping* and to be in a state of perpetual unsettlement. It is also true that differentiality in actuality holds a "minor" narrative back from being considered as a "tool" rather than as a "weapon". Micropolitical differentiation of a "minor" narrative is invested with "the direction", "the vector", "the model", "the expression" and "the desiring tonality" (Deleuze and Guattari 1987, 402) and thus it leads a "minor" text to function more as a "weapon" than as a mere "tool". Put in other words, a "minor" narrative can be considered as an effective "weapon" both to call rigid political segmentations into question and to reconstruct political segmentations in fluid combinations inasmuch as it is grounded in the logic of differentiality.

Theory of difference has been taken up both by Gilles Deleuze and Jacques Derrida in *Difference and Repetition* and *Writing and Difference* respectively. Deleuze attempts to understand difference in relation to repetition and speaks for the "conceptual difference" (Deleuze 1994, 13) in the act of repetition, whereas Derrida puts forward the concept of "différance" (1978, 161) to spell out the intricate relationship between text and meaning. Both of these theoretical frameworks inform the paradigm of differentiality in this way that the constituents of differentiality are soaked in both "deference" and "difference" so much so that it keeps up the molecular unfolding of a "minor" text intact. In addition to it, it also accounts for the unfinalizability of a "minor" text which always pursues the "lines of becoming".

Exteriorizing *Minor*: Topology, Assemblage and Immanence

It is undeniably true that "minor" has a distinctive topology which needs to be brought out to lay bare how assemblage, immanence and deterritorialization work hand in hand to make "minor" a zone of irreducibilities, indiscernibilities and inexorabilities. Topology of "minor" is marked out by contiguity, deformability and compossibility. In a "minor" space, entities catch up each other by the logic of contiguity—which, being grouped together with deformability and compossibility account for the "collective assemblage of enunciation". At this point, one may stop and think: what is an assemblage? How does an assemblage work? In response to these pertinent questions, one may straightaway refer to Deleuze and Guattari's critical viewpoint in *A Thousand Plateaus* where they reflect: "[an] assemblage, in its multiplicity, necessarily acts on semiotic flows, material flows, and social flows simultaneously" (1987, 22–23). What it suggests is that an assemblage performs like a multiplicity on the functional level and refers to the contiguous and continuous operations of a number of entities. In an assemblage,[3] one may hardly trace any order and organization as such and this is precisely the reason why an assemblage comprises "lines of escape", allowing its constituents to slide into a state of deterritorialization. Assemblage is apparently of two types—collective assemblage of enunciation and machinic assemblage—and interestingly, machinic assemblage resides in the collective assemblage of enunciation thereby forming a plane of immanence that stands tantamount to a multiplicity. In this regard, one may be reminded of Ian Buchanan's seminal work *Assemblage Theory and Method* where it is argued that it is "a point of departure" (2021, 13) that makes an assemblage stand out. It suggests that an assemblage is not merely an "arrangement" which is implicative of definitive order and sequence. An assemblage is ontologically bereft of strata and codes and is rather made up of "points" and "lines" and therefore any attempt to define it in territorial terms is bound to end up in territorialization which stands contrary to the fundaments of an assemblage.

Immanence, better to say, the "Plane of Immanence" connotes how the process of deterritorialization ends up in forming a "plane of consistency". The "Plane of Immanence" is constitutive of connections and disconnections, interiority and exteriority. Deleuze and Guattari pertinently observe in *A Thousand Plateaus*: "interior and exterior are equally a part of the immanence in which they have fused" (1987, 156). It means that the "Plane of Immanence" is divested of the tension between "inside and outside" and functions as a deterritorialized realm. In *What Is Philosophy?* Deleuze and Guattari have elucidated the "Plane of Immanence" as "the single wave that rolls them up and unrolls them. The lane envelops infinite movements that pass back and forth through it" (1994, 36). It connotes that the "Plane of Immanence" comprises co-extensive becoming of "intensities" which account for the actual functioning of "lines of escape". This idea is immensely useful for explaining how "minor" ends up forging a "Plane of Immanence" while making transgression through segmentarities. Just as the "Plane of Immanence" cannot be figured out in terms of structures and strictures, so also the "minor" can hardly be understood by means of rigid segmentarity. It is obvious that the "Plane of Immanence" caters functional logistical support to "minor" when it traverses "lines of escape". It also lays down a nuanced and complex network of connective synthesis sans externality thereby allowing "minor" to be in a dialogue with its own differential self and to give in the process of earth(ing).

Minor: An Effective Multiplicity

> *The concept of becoming is inseparable from that of minor literature, in that minor literature's deterritorialization of language necessarily entails a dissolution of cultural code.*
>
> ("Minor Writing and Minor Literature", Bogue 1997, 109)

In *A Thousand Plateaus*, Deleuze and Guattari have elaborately spelt out nuances of multiplicity—a terrain of "pure intensities" and "lines of escape". Multiplicity is generally understood as a conglomerate of "pure intensities" which ceaselessly traverse "lines of escape" to evade the process of territorialization. Deleuze and Guattari pertinently reflect: "[a] multiplicity has neither subject nor object, only determinations, magnitudes, and dimensions that cannot increase in number without the multiplicity changing in nature" (1987, 8). It means that in a multiplicity, one can hardly find any rigid segmentarity, pattern, configuration, homogeneity, and fixed point of reference. It can rather be figured out in terms of "determination", "magnitude" and "dimension" thereby corresponding to the features of a topological space. The terrain of multiplicity has many entries and exits and is therefore porous and polymorphous in nature. In this regard, it needs to be clearly mentioned that unlike diversity that refers to "many" in number, multiplicity connotes "many" in one and thus bears congruity to the mechanism of a BwO. Besides, multiplicity does not have either a beginning or an ending but a middle and is marked by the intersecting trajectories of irreducibility, indiscernibility and inexorability.

Connections between "minor" and multiplicity are densely nuanced and intersectional in nature. For example, as "minor" seeks to exceed the limits of territorialization and allows its constitutive functors to get engaged in a free interplay between mobility and mutability, supple segmentarity and fluid referentiality, one may also find a similar kind of free interplay between intensities, segmentarities and referentialities in a multiplicity. Like a multiplicity, "minor" can hardly be understood in terms of structured epistemology, for it is fluid and transgressional in nature. An understanding of "minor" is possible only if one takes its "determination", "magnitude" and "dimension" into account. It means that "minor" bears an exteriority of its own, which has definite and recognizable "determination", "magnitude" and "dimension". But, interestingly, the *intensive becoming* or differential intensity of "minor" does not comply with the laws of stratification and hence seems revolutionary in nature. Put in other words, it is because of the "quantum" flows of intensities, "minor" turns out to be one of the best political sites for reconstruction, experimentation and interrogation. This may lead one to argue that inherent "multiplicity" of "minor" helps itself work "effectively" at praxis.

"Effective" working of "minor" entails an irreducible yet recognizable disregard of the systematic immurement consisting of strata and codes. Effectiveness of "minor" is embedded in its multiplicity in the sense that while multiplicity leaves behind "lines of departure" and "lines of rupture" while getting intensified in compliance with spatio-temporal movements, "minor" capitalizes its innate multiplicity to get over the onslaughts of structurality and rigidity, thereby celebrating the potentialities of "singular-plurality".[4] In addition to it, one may find the "effective" working of "minor" in its ability to be in tandem with the principles of transgressionality, contingency and contrapuntality. In a nutshell, trajectories of irreducibility, indiscernibility and inexorability facilitate "effective" functioning of "minor" through the materialization of its multiplicity. Here, one may be reminded of Ken Gale's insightful intervention into "minor literature" in "Writing Minor Literature: Working With Flows, Intensities and the Welcome of the Unknown". Gale reflects how "minor" literature seeks to "de-centre and deterritorialize the dominance of 'major literatures' through strategies of experimentation, mistrust of traditional idioms and forms and of nurturing collective action" (2016, 304). Taking recourse to Gale's viewpoint, one may plausibly put forward that "effective" working of "minor" literature actually lies in the acts of "decentralization" and "deterritorialization"—the foundational pillars of multiplicity.

Notes

1 One may think of taking into account *Multiplicity and Ontology in Deleuze and Badiou* by Becky Vartabedian, where the author builds up a comparative analysis of "multiplicity" (2018, 182) between Deleuze and Badiou. In this work, she actually engages herself in exploring the following question: "what do we mean when we talk about 'being'?" (2018, 182).

2 The notion of "plasticity" is elaborately dealt with in *The Future of Hegel* by Catherine Malabou who comments on the destructive potentials of plasticity while spelling out its dynamicity. Whereas in *The Future of Hegel*, Malabou resorts to "dialectical process" (2005, 25) to explain how plasticity can be realized, she seems to have gone beyond the limits of dialecticality in *Ontology of the Accident* to underscore the manifestations of "destructive plasticity" (2009, 11) in reality. Ghosh seems to have taken epistemological cues from Malabou's formulation.

3   In *Artmachines: Deleuze, Guattari, Simondon*, Anne Sauvagnargues has examined six important aspects of machinic assemblage. While elucidating nuanced dimensions of machinic assemblage, Sauvagnargues reflects: "the machine only exists in the plural, as the cutting of flows, a machine that cuts a flow in turn gets cut by other machines" (2016, 190). It suggests that machinic assemblage consist of cuts and folds and aims at eradicating the possibility of any forceful attribution of meaning. In a way, it levels down all forms of hierarchy and stands rooted in the dynamics of deterritorialization.

4   Jean-Luc Nancy has propounded the concept of "singular-plural" in the following terms: "A singularity is always a body and all bodies are singularities" (2000, 18). Nancy means to say that the existence of singularity always stands "with" other beings and thus a singularity is bound to be understood as a "singular-plural".

III Cartography of Blue Humanities
Contentions and Contestations

Eating the ocean: We do it every day, often without knowing it. Humans have eaten the ocean for as long as we've been around ... Now we are at risk of eating it up.

(*Eating the Ocean*, Probyn 2016, 2)

The Sea is history.

("The Sea Is History", Walcott 1986, 354)

Renowned critical thinker Rachel Carson pertinently holds in *The Sea Around Us* that the human world is not just surrounded by the ocean but also the ocean stands as an immanent living repository of human and non-human "excesses" that can provide epistemic breakthroughs to figure out the "creation of life from non-life" (1961, 3), whereas Sidney I. Dobrin moves a step forward and reflects on how the ocean nowadays occupies a "secondary" position in different ecocritical projects in *Blue Ecocriticism and the Oceanic Imperative*: "ocean has remained a secondary concern in the ecocritical project, a second class citizen of the ecocritical domain, despite its ever presence in all matters terrestrial" (2021, 227). Following Dobrin's critical concerns, this chapter seeks to put the spotlight on the nuanced and seamless overlapping between the oceanity of the ocean and coastality[1] of the coast in general and particularly aims at remapping the frayed epistemic strands of Blue Humanities, taking into account its specific platitudes and latitudes.

Blue Humanities: Contexts and Concerns

the development of marine or "blue" humanities calls for enmeshment between cultural history (traditionally the domain of the humanities) and natural history (aligned with the sciences) ("Widening Gyre" Bloomfield 2019, 503–504)

. . . the deep seabed is simultaneously understood as a resource frontier, framing both salvage and seabed mining vis-à-vis settlement frontier imaginaries. ("The Blue Frontier" Han 2019, 463)

DOI: 10.4324/9781032629728-4

Do oceans exist at the end of the human world? Or, does human world start off the oceans? The entanglements between the human world and the marine world are of manifold significance simply because neither of them can be properly understood in isolation nor can they exist in isolation. Connections between terrestrial world and marine world are deeply nuanced and marked out by porosity, non-linearity and contiguity, and hence are quite problematic. Besides, much of the depth of the marine world still remains unexplored to the human world—an induced inquiry into the "being" of the ocean thus can help researchers and scholars work out fresh insights for ungrounding transformative becomings of oceans. For instance, continual (form)ation and changing form(ability) of oceans are of much critical interest to scholars hailing from the field of Oceanic Studies inasmuch as "differentiality" and "processuality" play instrumental roles in conditioning the ceaseless trans(form)ation of oceans circumstanced by anthropocentric overtures. Interface between the human world and the marine world has already been approached from different vantage points including Critical Ocean Studies, Oceanography, Oceanic Studies, Oceanic Engineering, Cultural Studies, Transoceanic Studies, Diaspora Studies, Marine Geography, Marine Biology, and so on, but unfortunately, no critical school of thought is able to provide effective and integrated ways of looking into the layered entanglements and disentanglements between the human world and the marine world as of now. Moreover, an ocean has a lot of interconnecting ecosystems—the complex functioning of which makes an ocean stand in non-compliance with rigidity, structurality and fixity. Sometimes, irresponsible and insensible people throw away different kinds of waste into oceans other than the continual disposal of industrial wastes into the oceans, resulting in the endangerment of marine species, acidification of marine water, affection of entwined marine ecosystems and imperilment of the lives of coastal locals, among others.[2] These material contexts and concerns lay down a solid platform for Blue Humanities to rise as an interdisciplinary area of study which aims at making oceans figure in the human consciousness by means of redrawing the human-ocean interface.

In a seminal article "Toward a Blue Humanity?", Ian Buchanan and Celina Jeffery clearly put forward that at a moment when oceanic resources are rapidly depleting and disappearing, there comes up a "growing body of work" called "Blue Humanities" which seeks to " historicizing the ocean and making it part of contemporary consciousness in a way" (2019, 12). Blue humanists, in their view, look forward to exploring a sort of "politics of invisibility at work here" (2019, 11) and *saving* the oceans from ecological catastrophes and human exploitations. Blue humanists can, for that matter, consider oceans as "missing contexts" (Buchanan and Jeffery 2019, 11) while attempting to figure out appalling environmental impacts on locals. At this point, it needs to be made clear that unlike ecological studies, Blue Humanities stands as an interdisciplinary epistemological framework which seeks to (re)orient the human world to the marine world that stands at jeopardy in the age of Anthropocene.

Stacy Alaimo, an important critical thinker in the field of Blue Humanities, goes a step further in "Introduction: Science Studies and the Blue Humanities" and observes that blue humanists aim at developing a kind of "environmentally oriented scholarship" that facilitates them to get critically engaged with "epistemological problems of scale, onto-epistemologies of rapidly altering and utterly entangled lifeworlds, and the urgency of extinction" (2019, 431). Taking recourse to this pertinent observation of Alaimo, one may try to understand Blue Humanities in this way that apart from examining oceans physically, blue humanists are encouraged to build up a distinct epistemology to deal with the "differential" progressions of oceans, complex and entwined marine ecosystems, and the rapid depletion of marine species.

Critical questions concerning "scale, temporality, materiality, mediation" (Alaimo 2019, 431) are of profound importance inasmuch as these can help blue humanists map oceanic transfigurations. In a way, Blue Humanities seeks to put "oceans at the forefront of our attention", intending to posit "cultural history" in a blue context ("Shakespeare and Blue Humanities", Mentz 2019b, 384). Oceans possess disparate histories that are very much connected to cultural changes. Put in other words, oceans have been exploited by sailors, colonizers, fishermen, pirates, and so on for different purposes. Therefore, instead of looking at oceans as a *cluster of dead zones*, blue humanists are pushed to take into consideration "material" dimensions of oceans—the becoming of which stands in tandem with cultural, political, and economical deterritorializations and reterritorializations.

Thus, in "The Prospect of Oceanic Studies", Hester Blum emphatically argues: "The sea is not a metaphor" (2010, 670). Based on Blum's timely intervention into the material dimension of oceans, one may cogently moot that "the material conditions and praxis of the maritime world" (2010, 670) bear enormous significance in the context of Blue Humanities in that the physical existence of the marine world reflectively mirrors the happenings in the human world, and most importantly, documents vicious anthropocentric overtures at it.[3] Blue humanists' strong leanings to the onto-epistemological transformations of oceans call figurative understandings of oceans into question, for oceans stand as a "more nuanced than imagined reality" (Smith and Mentz 2020, 1). In a seminal article "Learning an Inclusive Blue Humanities: Oceania and Academia through the Lens of Cinema", James L. Smith and Steve Mentz have made attempts to widen the scope of Blue Humanities by contending that it is actually a "critical practice" which "runs an ongoing risk of being co-opted by imperial maritime histories, racializing ideologies, and the interests of capitalism" (2020, 1). This observation seems quite engaging not only because it takes into account rhizomatic flows of capitalism while figuring out constitutive epistemological strands of Blue Humanities but also because it makes Blue Humanities amount to a "critical practice" that is instrumental in spelling out *differential intensities*[4] of oceans. *Differential intensities* of oceans cannot be comprehended without resorting to the free flow of energy through complex and entwined marine ecosystems. It thus pushes blue humanists to take into consideration how "the oceanic flows of capital and energy" (Scott 2020, 1) condition the transformations of oceans in particular.

There are some critical terminologies—"transoceanic imaginary" (DeLoughrey 2007, xi), "wet ontology" (Steinberg and Peters 2015, 2), "metageography" (Lewis and Wigen 1997, 189), and so on, which help one critically engage with the oceans. But, these do not seek to foreground material transformations of an ocean, which stands as one of the important concerns of Blue Humanities. Apart from this, blue humanists could think of formulating a blue epistemology which is crucial for mapping ontological and ontical alterations which oceans undergo.

Configuration of a blue epistemology is therefore intended to be designed in such a way that it can help critical thinkers track down ways of exploiting marine resources and develop a reorientation in the minds of common folk to oceans.

Blue humanists are mysteriously silent on the (inter)connections between marine and terrestrial ecologies—coastal areas which give room to illicit trading and deposition of human and industrial wastes. In other words, margins of the marine world are of profound importance, for the continual damaging of coastal ecologies by means of dumping toxic and plastic wastes[5] entails catastrophic impacts on the marine ecosystems—which leaves the planet Earth in great distress. Blue humanists need to exercise activism and radicalism not only in terms of working out an advanced blue epistemology but also in terms of practising exemplary roles in putting rules in place to prohibit human violence against oceans. The epistemological limit of Blue Humanities stands inclusive of its proclivity to emerge as a meta-narrative, thereby not paying an adequate amount of attention to the local issues and problems. It emerges as an elitist epistemic framework that discounts the sufferings and miseries of human and non-human beings residing in coastal regions. Besides, Blue Humanities offers a densely hierarchized and stratified epistemic model of thinking which does not help one understand the interface between oceans and humans.

Thus, systematic expansion of Blue Humanities has to be done to develop an integrated and grounded epistemology to examine both "intensive" and "extensive" movements of oceans. In addition, integration of coastal ecological specificities with the existing grammatology of Blue Humanities is intended to pull the latter back from being labelled as a "meta-narrative".

Blue humanists need to ponder over geo-centric philosophies like geo-philosophy and "gaia philosophy", among others, to expose how the wellbeing of oceans stands in tandem with sustainable development of the Earth. Blue humanists also need to focus on "blue archive" to dig out geo-historical inscriptions lying under the marine world in order to make oceans figure in the anthropological consciousness of marine ecology. Another important concern of Blue Humanities has to be the production of a strong counter-discursive and resistive framework to safeguard marine resources from depletion and extinction caused by the pervasive forces of "energopolitics".

Blue Archive: Remnants of *Blue Trauma* and *Blue Memory*

. . . the story of trauma is inescapably bound to a referential return. (Unclaimed Experiences: Trauma, Narrative and History 1996, Caruth, 7)

At this point, one may stop and mull over a couple of interesting questions: What exactly is a *blue archive*? How do *blue trauma* and *blue memory* contribute to the ceaseless de-structuring and re-structuring of a *blue archive* in alignment with spatio-temporal alterations? In order to critically engage with the formation of a *blue archive*, one has to take recourse to *Archive Fever: A Freudian Impression*, where Jacques Derrida lays down onto-epistemological dimensions of an "archive" while pointing at its enormous potentials. An archive happens to be a site for both "*commencement*" and "*commandment*" (1995, 9)—a place from which "physical, historical or ontological" movements start off on the one hand and on the other hand, a place for the exercises of "authority" and "social order" (1995, 9).[6] Following Derrida's viewpoint on the formation of an archive, it can cogently be argued that "archive formation" is a sort of political exercise executed to actualize the dialectical interplay between the politics of inclusion and exclusion. So far as the ratification of the authoritative potentials of an archive is concerned, Derrida opines that there can be no archive without "a place of consignation", "a technique of repetition" and "a certain exteriority" (1995, 14). He means to say that an archive always occupies a topographical territory, operates by the logic of repetition and essentially has an exteriority of its own. Holding an archive "hypomnesic" in nature, Derrida puts forward that an archive could be held as a *material memory* of something written, performed or heard.

Whereas Derrida's notion of "archive formation" is fraught with temporal, topographical, spatial and territorial limits, *blue archive* stands wedded to the "quanta of deterritorialization" and therefore gets continuously built up in terms of "flowing", "folding" and "circulating". *Blue archive* actually stands mediated through marine ecologies and is characterized by poroplasticity, processuality, and "trans(in)fusion". *Blue archive* can be comprehended as a borderless spatiality which comprises "excesses" of human wastes, industrial untreated sewages, toxic chemicals, indestructible plastics, dead bodies of endangered species, remains of human naval architectures, remnants of human cultural histories, wreckages of human engineering materials, and so on.

Interestingly, the formation of *blue archive* does not stand subjected to anthropocentric manipulations and is rather contingent upon "geokinesis".

Blue archive actually works against itself, showing denials and refusals to any sort of spatio-temporal territorialization.

Another fascinating dimension of *blue archive* is that one cannot return to the configuration of *blue archive* twice inasmuch as it is potentially charged with what Deleuze and Guattari call "lines of becoming" in *A Thousand Plateaus*. In other words, the workings of *blue archive* stand in tune with the intensive and extensive becomings of oceans. Here, one may be reminded of Donna Honarpisheh's reflective article "The Sea as Archive: Impressions of *Qui Se Souvient De La Mer*" in which she cogently argues that the sea could be held as a:

> liquid materiality whose continuous undulation both erases and re-generates ... To think of the sea as an archive is at once to think (in a Derridean sense) with the place of the archive as home, domicile, and residence, and in a second register to point to those extra-juridical figures, rhythms, or gestures that escape or remain outside the purview of hermeneutic power.
>
> (2019, 93–98)

Taking recourse to Honarpisheh's timely reflection on the archival potentials of the sea, one may contend that *blue archive* works by the subtle interplay between erasure and re-generation, addition and deletion, striation and smooth(ing), territorialization and trans-territorialization. Apart from that, *blue archive* is elusive in nature in the sense that the distinct (trans)formation and (trans)form(ability) of *blue archive* makes it/itself a (trans)positional entity and therefore, it bears revolutionary potentials in it. It is argued here that it is because of the *differential intensities* of an ocean that *blue archive* continually de-forms and re-forms itself and always stands in a state of flux and "trans-".

Nuanced operativity of *blue archive* can further be worked out by means of taking into consideration *blue trauma*[7] and *blue memory*.[8] The notion of *blue trauma* can be configured as a continual and impactful precipitation of the dire consequences of violent human activities in the seabed, in the garb of "trauma". Generally, vicious and violent human activities carried out either in the midst of a marine world or on the coastal locations result in the *blue trauma*—the location of which ceaselessly shifts from the margin to the centre and vice versa. Conceptualization of *blue trauma* actually stands grounded in the notion of "ocean trauma":

> because of man's mistreatment of natural resources on land, erections of giant oil-rigs at sea and great numbers of ships traversing the world's oceans, sediments have also come to contain many harmful substances such as heavy metals, oil spills and pollutants, discarded fishing nets,

sewerage and garbage resulting in ocean trauma ... All this misuse and abuse of sea life and habitat is, in effect, ocean trauma.

(Discover Sedgefield 2022, n.pag.)

It thus becomes clear that *blue trauma*, in fact, speaks to the intense traumatization of the marine world caused by detrimental human activities across coastal and marine territories.

Epistemic understanding of *blue trauma* followed by its distinctive figurations is supported by strands of *blue memory*. One may question: What is the expression of *blue trauma* in reality? Does *blue trauma* have any connection with *blue memory*? So far as the figuration of *blue trauma* is concerned, one may be reminded of the notion of "catastrophism",[9] as it has been understood by Bronislaw Szerszynski and John Urry in "Changing Climates: Introduction": "[catastrophism happens to be a] framework which emphasizes non-linearity, thresholds and abrupt and sudden change" (2010, 2). This can offer some insights to look into the material reflections of trauma across marine ecologies. Considering epistemic strands of "catastrophism", one may tenably put forward that *blue trauma* finds material figuration in the form of ecological catastrophe which is the concomitant upshot of neoliberal human exploitation of marine ecology. *Blue trauma* actually gets deepened in a "non-linear" fashion and takes "sudden change" in terms of movement and manifestation. Besides, the dynamics of *blue memory* is grounded in the *blue network* that refers to the interconnected ecologies in the marine world and its nuanced connections with terrestrial counterparts. *Blue memory* therefore can be comprehended as a "portal" to access the territory of the marine world which could be held as a "marine repository" comprising the remnants of human neoliberal exercises. Epistemic formulation of *blue memory* stands wedded to what Fred Wang argued in "Ocean Memory: Humans Have Memories, Oceans Can Have Them Too":

Human beings have memories ... Ocean can have memory as well. It is a huge bank, and stories tremendous amount of information in it, in a variety of different aspects . . . many discoveries can be found in the ocean through its memory, in various different forms. This exciting idea of comprising the ocean as a whole, and to learn as much things as possible about the world from the information provided by the ocean, is the idea of ocean memory The ocean has an integrated memory that comprises of nearly everything from the birth of the earth.

(2021, n. pag.)

Following this contention, it can be put forward that the materialization of *blue memory* entails the collection of the consequences of unrestrained human atrocities to the marine world.

Therefore, *blue memory* can be worked out as a critical way to make inroads into the realm of *blue trauma* in order to comprehend how oppressive and neoliberal frameworks put *blue archive* at stake.

Remapping Blue Humanities: Instrumentalizing Coastal Ecology

. . . the ocean as a vital starting place to develop what I call milieu-specific analysis, calling attention to the differences between perceptual environments and how we think within and through them as embodied observers. (Wild Blue Media Jue 2020, 3)

The core intellectual challenge of the Blue Humanities explores explores how water functions in and across multiple scales. (An Introduction to Blue Humanities Mentz 2014, xiii-xiv)

. . . a "Critical Ocean Studies" that flows across disciplines; dives into submarine depths and submersions; swims into multispecies entanglements; intersects with feminist, indigenous, and diasporic epistemologies; recognizes the agency of a warming, rising ocean; and transforms our critical inquiries and reading practices. ("The Ocean in Us" Perez 2020, 2)

Remapping Blue Humanities in terms of considering coastal ecological concerns entails epistemic deterritorialization of Blue Humanities so as to help it better address ecological threats to both the oceans and coasts. It is argued that a cartographical mapping of coastal ecology is of profound importance to the act of remapping Blue Humanities inasmuch as coastal ecology is a terraqueous territory which at once receives throwaways of human beings and at times lets them get engulfed by the sea, and therefore any disruptive and destructive activity in coastal ecology is bound to get precipitated in the *blue archive*. In order to elucidate the epistemic remapping of Blue Humanities, the notion of "saturation thinking"—a sort of elemental politics which can cater logistic support to Blue Humanities in explaining the trans(in)fusive becomings and "mediatedness" of the ocean through material entities—needs to be taken into account. In "Thinking with Saturation Beyond Water: Threshold, Phase Change and The Precipitate", Melody Jue and Rafico Ruiz attempt to lay down "saturation thinking" in terms of "threshold", "phase change" and "the precipitate" to acknowledge the "co-presence of multiple phenomena within the complexity of our symbiogenic world" (2021, 3). They reflect:

> Saturation is not only about elements in political configurations, but about the politics of elemental formation through their mediations We think of elementality not as a taxonomy of substances, but as a politics of co-presences under flux Saturation embodies the fluidity of relations that exceed attempts to contain, manage, and fix them to stable frameworks.
>
> (2021, 4–7).

Following the insights of Jue and Ruiz, one may argue that the ocean is a densely mediated and interconnected materiality which makes differential and co-intensive progression through its constitutive elements. The ocean is at once a medium and at times mediates "elemental formation" across porous material bodies. Linking "saturation thinking" with the act of remapping Blue Humanities is intended to examine the "co-presences" of the fluid materialities in the "symbiogenic world". Put in other words, the operative logic of remapped Blue Humanities bears "blue (infra)structural"[10] underpinnings, thereby suggesting how *blue archive* stands caught up in the "lines of becoming". One may refer to "'A Perfect and Absolute Blank': Human Geographies of Water Worlds" where Jon Anderson and Kimberley Peters cogently reflect:

> we can accept that the sea (and even the world in general) is not static or stable, it is only of the immediate present, before it becomes something else. It is on a line of mobility and flow that it constantly taking it elsewhere.
>
> (2014, 12).

This critical reflection accounts for the ontical and ontological processuality of *blue archive*.

At this critical juncture, one may stop and mull over some intriguing questions relating to the effectuality of remapped Blue Humanities in connection with the energopolitical exploitation of blue species: How does neoliberal framework put blue species at stake? How does remapped Blue Humanities work in the context of the endangerment of blue species? How does remapped Blue Humanities work against the dynamics of "energopolitics"?

It is true that the neoliberal framework sets the stage ready for the endangerment of blue species and at times leaves vicious impinges on the multispecies interactions. In the neoliberal regime, ruling authority and affluent entrepreneurs configure "blue economy"[11] as a set of governing principles by means of which they pounce upon marine resources, intending to export and exoticize rare and dying out marine resources. Whereas Amitabh Kant, Pramit Dash and Piyush Prakash refer to "blue economy" in "Leapfrogging the Indian Blue Economy" in the following terms: "Blue Economy in general is built on some key building blocks that propel economic value creation, foster sustainable livelihood and restore and maintain ocean ecosystem." (Employment News, 12 March, 2022), Young Rae Choi prefers to figure it out in "The Blue Economy as Governmentality and the Making of New Spatial Rationalities" in a different fashion: "Blue Economy practices of seeking economic ways to use space ironically leads to the representation of sea space as potential development space and eventually to more intensive and extractive uses of sea space as a consequence" (2017, 39). Following the two different observations, one may reflectively comment that "blue economy" stands for a set of flexible governing ideologies which help one exact and extract economic gains out of natural marine resources. Materialization of "blue economy" differs in varied topographical locations, depending on the availability and accessibility of marine resources and their demands in the (g)local markets. Crass and manipulative materialization of "blue economy" actually results in the endangerment of blue species including coral reefs. In "Integrated Ocean Management for a Sustainable Ocean Economy", Jan-Gunnar Winther et al. cogently reflect: "[the] evolving ocean economy, driven by human needs for food, energy, transportation and recreation, has led to unprecedented pressures on the ocean that are further amplified by climate change, loss of biodiversity and pollution" (2020, 1451) Winther et al. rightly observe that reckless exercises of "ocean economy" entails "loss" of several marine species and augmentation of marine pollution.

"Blue economy" can be held as a state-ratified exploitative framework which makes room for *blue trafficking* along coastal regions. *Blue trafficking*, a derivative of "blue economy", refers to the illicit and (g)local trafficking of marine resources. Based on the demands in the market, *blue traffickers* engage locals in extracting natural resources from under the marine world and then in selling these items to them so that they can export it (g)locally. One interesting aspect of *blue trafficking* is that instead of exploiting human resources directly, marine resources are precisely targeted here because on the one hand, concentration of marine resources along the coasts is getting reduced due to overall marine pollution and thus coastal-marine resources have high economic value in the market and on the other hand, it is not always possible to make inroads into the marine world for those who stay away from coastal regions, hence coastal marine resources have extraordinary capital value for neoliberal elites who live away from the coastal regions. *Blue trafficking* steadily disseminates along coastal regions because of the impoverished conditions of locals dwelling in coastal areas. It provides a kind of temporary employment to those who are in dire need of money and strive to withstand the buffets of neoliberal capitalism. Thus, it has become quite clear that endangerment of blue species does not remain confined in the marine world; rather, it gets permeated across terrestrial ecologies by means of *blue trafficking* which runs by the logic of *neoliberal dispossession* of oceans, and thus, in a nutshell, is a vicious neoliberal capitalist enterprise. Considering the rapid dissemination of *blue trafficking*, one may tenably argue that the epistemic remapping of Blue Humanities stands capable of providing a counter-discursive framework so as to impede the operation of *blue trafficking* conditioned by the pervasive forces of "energopolitics". Besides, along with addressing potential dangers pertaining to the wellbeing of the oceanic ecologies, remapped Blue Humanities seeks to put the spotlight on the sufferings and miseries of naive locals who depend on local markets financially and cannot often resist the temptation of being rapacious in a neoliberal world. It serves to accentuate the instrumentalization of coastal ecology in taking on the onslaughts of *blue trafficking* which, in many ways, stands disruptive to the wellbeing of the oceans. It is therefore argued that a remapped blue humanitarian framework has to take note of coastal ecological concerns including the corrosion and erosion of coastal regions, irregular cleaning up of dirt and wastes from coastal areas, illicit trafficking of blue species and illicit marine constructions for smooth functioning of neoliberal economy, among others, so to ensure seamless interactions and intersections between the oceanity of the ocean and coastality of the coast.

Remapped Blue Humanities is thus claimed to be able to upset and unsettle energopolitical exploitations of marine and coastal resources. For instance, blue humanists are encouraged to make use of the interlinkages among *blue archive, blue trauma,* and *blue memory* embedded in the nuanced intersections between the oceanity of the ocean and coastality of the coast to interrogate how neoliberal elites execute energopolitical frameworks to make much of the "blue economy".[12] In a nutshell, it is by enfolding coastal ecological concerns in the analysis of the potential dangers that are threatening marine life at large, that statist, exploitative and provincial mechanisms of "energopolitics" reigning in the intersections between the oceanity and coastality can be interrogated with the help of remapped Blue Humanities which ultimately renders "energopolitics" dysfunctional and unproductive.

Notes

1 In "Storied Seas and Living Metaphors in the Blue Humanities", Serpil Oppermann pertinently observes that an ocean is always in a state of flux, and thus creates and re-creates a plethora of possibilities at figurative, epistemological and material levels. Oppermann shows her deep concerns for the "biogeophysical existence" (2019, 443) of an ocean which is often understood in figurative terms only. She draws one's attention to the "material and discursive contexts" of seas while propounding "material ecocritical" theory (2019, 445). Additionally, in "The ocean exceeded:Fish, flows and forces", Christopher Bear speaks of how the oceans "exceed their material, discursive and imagined boundaries along with their liquid form" (2019, 329).

2 In the "Introduction" to *Coastal Change, Ocean Conversation and Resilient Communities,* Marcha Johnson draws attention to the fact that coastal culture nowadays faces the challenge of finding out "a way to live on shorelines that embraces, celebrates, and supports a healthy ocean" and "healthy networks of ecosystems that allow the ocean to thrive" (2016, 1). It is thus quite clear that coasts are of immense interest to blue humanists who are concerned with the marine world.

3 Philip E. Steinberg is one important thinker who has also brought out the significance of fluidity and mobility of the marine world in "Of Other Seas: Metaphors and Materialities in Maritime Regions". Steinberg argues that "connections and flows" (2013, 156) make oceans stand as "more-than-human assemblage" (2013, 156).

4 *Differential intensity* refers to the intensive becoming of an entity. It suggests how an entity differentially progresses through spatio-temporalities. Deleuze and Guattari have explained it at length in *A Thousand Plateaus.*

5 In *Mythologies,* Roland Barthes understands "plastic" as "the very idea of its infinite transformation" (1957, 97), thereby suggesting its infinite and wonderful malleability. Catherine Malabou takes up the concept of plasticity in *The Future of Hegel* and holds it as a "receiving and a giving of form" (2005, 84). On the other hand, in "Plastic Literature", Ranjan Ghosh has elaborated on how "plastic literature" offers a new way of thinking about and doing literature: "plastic provokes connections/comparatism with a difference that speaks of becomings, dispersions,

and immanence . . . "plastic lit(t)erization" [seeks] to bring about a new poetics of thinking and doing literature" (2019, 277). In "The Plastic Controversy", Ranjan Ghosh maps different dimensions of plastics—"surprise", "bizarre" and "becoming"—while commenting on the vicious impacts of plastics on the Earth.

6 In "The Power of the Archive and its Limits", Achille Mbembe puts forward the concept of "becoming an archive" (2002, 19) apart from drawing some onto-epistemological limitations to it. This article also seeks to explain religious and architectural dimensions of an archive.

7 The colour "blue" has nothing to do with trauma or the process of traumatization. Rather, it only signifies the marine world and its trauma. One may also find a complex bond among "trauma", "memory" and "history" well-elucidated in *Unclaimed Experiences* by Cathy Caruth.

8 Here, "blue" in *blue memory* represents the marine world. One may find interest in reading Andrew Jones's *Memory and Material Culture* to know about the nuanced operation of memory in compliance with "remembering" and "forgetting" (2007, 31)

9 In *Catastrophe Theory*, V.I. Arnold theorizes catastrophe in the following terms: "*Catastrophes* are abrupt changes arising as a sudden response of a system to a smooth change in external conditions" (1992, 2). Arnold means to say that catastrophe is steeped in singularity and is capable of unsettling usual functions of a system.

10 In *Blue Infrastructures: Natural History, Political Ecology and Urban Development in Kolkata*, Jenia Mukherjee elucidates: "[Blue infrastructures] ... is a story comprising the interplay between complex narratives of continuous functioning of the natural and the cultural, the physical and the manufactured, the tamed and the untamed, where these remain enmeshed as embedded entities, infusing meanings into this integrated scape" (2020, 3).

11 In "Actualizing Marine Policy Engagement", Jennifer Brewer takes up different aspects of "blue economy" and suggests that marine spaces need to be planned in such a way that they can cater to the needs of people belonging to different strata of society. Brewer proposes the use of "cartographic authority" to divulge the gaps between "fixed maps" and "shifting ocean currents" so that local issues can be sorted out (2017, 48).

12 In *Structures of Coastal Resilience*, Catherine Seavitt Nordenson, Guy Nordenson, and Julia Chapman think that coastal areas need to be built up and re-built in compliance with the shifting intensity of ecological catastrophes so that coastal towns and cities can be safeguarded. Instead of viewing marine disasters as "challenges" for the coastal areas, they need to be used to work out plans to augment "coastal resilience" that is required for sustaining the good health of coastal ecology (2018, 3).

IV Geokinetic Interventions into Matter and Matter(*ing*)

Thresholds of *Energopolitics*

> He explained that the Earth—the Deterritorialized, the Glacial, the giant
> Molecule—is a body without organs. This body without organs is permeated
> by unformed, unstable matters, by flows in all directions, by free intensities or
> nomadic singularities, by mad or transitory particles.
>
> (*A Thousand Plateaus: Capitalism and Schizophrenia*,
> Deleuze and Guattari 1987, 40)

> Thinking theory is thinking energy. Is energy really a thought? Or does it stay
> as "thinking'" in the process of uncovering a thought?
>
> (*Trans(in)fusion: Reflections for Critical Thinking*, Ghosh 2021, 43)

Energy is the essence of every substance. Everything—including mass and every material thing—is a form of energy. (Ayre 2016, 2)

In the postglobalized contexts, developed nations of the world have been in search of new energy resources to make up the energy deficiencies experienced in different fields of industry.[1] Energy is of profound importance not only because it causes circulation of economic activities across the world but because it is nowadays being turned into a political tool to exert power on developing nations which look forward to developed nations' supply of energy. It is true that energy resources stand scattered across the world and political attempts are being made to pounce upon those energy resources which could be capitalized to rule over the rest of the world. Having control on the production and dissemination of energy along the networks of economy becomes an important political move that developed nations take into account to refashion their political dominance over the rest of the world. Considering this emerging trend of politics in terms of energy, this chapter seeks to contend that the geology of "matter(ing)" has to be taken into account to figure out the *allagmatics* of energy through geological, political and cultural frameworks of the Earth. In a nutshell, Dominic Boyer's configuration of "energopolitics"[2] followed by its nuanced operation, in particular, are intended to be questioned by means of both putting it in a series of dialectical interactions with biopolitics from which the former has arguably come off and by taking recourse to the epistemological strands of remapped Blue Humanities.

DOI: 10.4324/9781032629728-5

Geology of Matter(ing): How Does Energy Matter?

. . . "energy" . . . is in fact a study of "energies", derived from technology, material culture, and intellectual culture in equal measure. ("Medieval Water Energies" Smith 2018, 1)

Free intensities" or "nomadic singularities" of "matter" may lead one to argue that the transductive and transformative potentials of energy can fully be grasped by means of particularly considering geology of matter(ing) which can help one delve deeper into the onto- epistemological dimensions of energy. Varied geo-centric approaches existing both in Indic and Western epistemological traditions are taken into account to spell out how the geology of matter(ing) plays an instrumental role in deterritorializing energy from the bounds of Anthropocentrism. For example, "informed" qualities of "matter" were duly acknowledged in Vedic times when planetary consciousness started to take shape. One may cite a couple of Vedic hymns to elaborate how the dynamic vibrancy of "matter" was understood in the ancient Indic tradition:

> I call for Aditi's unrivalled bounty, perfect, celestial, deathless, meet for worship. Produce this, ye Twain Worlds, for him who lauds you. Protect us, Heaven and Earth, from fearful danger.
>
> (1.185.3, *The Hymns of the Rigveda*, Griffith 1896, 99)

> Come hither, Agni; sit thee down as Hotar; be thou who never wast deceived our leader. May Heaven and Earth, the all-pervading, love thee: worship the Gods to win for us their favour.
>
> (1.76.2, *The Hymns of the Rigveda*, Griffith 1896, 41)

> How shall we pay oblation unto Agni? What hymn, Godloved, is said to him refulgent? Who, deathless, true to Law, mid men a herald, bringeth the Gods as best of sacrificers?
>
> (1.77.1, *The Hymns of the Rigveda*, Griffith 1896, 41)

These Vedic hymns at once celebrate the protective and caring actions of *dyāvāpṛthivī* and at times underline the onto-epistemic fluidity of matter that constitutes the material figurations of the Earth. What it implies is that in Vedic knowledge systems, through the celebration of the diverse functions of *dyāvāpṛthivī*, innate dynamicity of matter is recognized. In other words, it is the flow of the matter that causes the Earth to move forward. Whereas Indic understanding of the geology of matter(ing) lies in recognizing the vibrancy of matter, Western understanding of the same is more inclusive and attached to the "free intensities" or "nomadic singularities" of matter. Whereas Deleuze and Guattari propose to view the "materiality" of the Earth in terms of "geophilosophy" that speaks of the nuanced co-relationality between reterritorialization and deterritorialization as reflected in *What Is Philosophy?* Karen

Barad offers an engaging insight on the intersections among matter, energy and meaning in *Meeting the Universe Halfway*:

> Matter and meaning are not separate elements. They are inextricably fused together, and no event, no matter how energetic, can tear them asunder … Mattering is simultaneously a matter of substance and significance, most evidently perhaps when it is the nature of matter that is in question, when the smallest parts of matter are found to be capable of exploding deeply entrenched ideas and large cities.
>
> (2007, 3)

Following Barad's nuanced understanding of "matter", one may argue that Barad is particularly interested in the f(a)ct of mattering which refers to both "a matter of substance and significance". Induced by Barad's configuration of "mattering", one may put forward that the "mattering" of energy is even more densely nuanced, complex and intensive. "Mattering" of energy primarily entails the very (f)act of *affecting* something. It implies that "mattering" of energy results in the contestation of "fixity", "inclusivity" and "exclusivity". "Mattering" of energy can be marked out by "trans-plastic-habit"[3] which allows energy to be disengaged from any sort of either "fixed", "inclusive" or "exclusive" pattern. In order to grasp "mattering" of energy in epistemic terms, one may choose to reckon with dual analytics of "magma" and "ensemble". Put it in simple words, whereas "magma" refers to "more than a simple mix" (Ghosh 2021, 14), "ensemble" is understood as a "collective" of many. Whereas "magma" is steeped in the logic of indeterminacy, "ensemble" is soaked in determinacy. In "The Logic of Magmas and the Question of Autonomy", Cornelius Castoriadis attempts to define "magma" in relation to "ensemble" in the following terms:

> M1: If M is a magma, one can mark, in M, an indefinite number of ensembles.
> M2: If M is a magma, one can mark, in M, magmas other than M.
> M3: If M is a magma, M cannot be partitioned into magmas.
> M4: If M is a magma, every decomposition of M into ensembles leaves a magma as residue.
> M5: What is not a magma is an ensemble or is nothing.
>
> (1997, 297)

Castoriadis means to say that every disintegration of a magma results in "ensembles" which render magma as residue. Considering Castoriadis's formulation of "magma", it can be argued that "magma" cannot be defined because it is truly paradoxically magmatic . Similarly, *energy linkages*[4] can be understood as "magmatic", for temporary (en)settlement of energy into material bodies renders it *energy ensemble*, that is, a "collective" of energies. In a way, it seems that it is the "mattering" of energy that renders it *energy ensemble.*[5] Conversely, it is also true that *energy linkages* bear "magmatic" dimensions in it in the sense that it always refuses any sort of categorical territorialization and therefore cannot be held tantamount to an "ensemble". What it suggests is that it is the (f)act of "mattering" of energy that makes energy all the more problematic and subject to epistemic unfolding.

Jane Bennett is one important thinker who has exclusively dealt with the transformative vibrancy of matter in *Vibrant Matter: A Political Ecology of Things*, which attempts to connect the forces of vibrant matter with deterritorial and fluid operativity of an assemblage.

> Assemblages are ad hoc groupings of diverse elements, of vibrant materials of all sorts. Assemblages are living, throbbing confederations that are able to function despite the persistent presence of energies that confound them from within. They have uneven topographies because some of the points at which the various affects and bodies cross paths are more heavily trafficked than others, and so power is not distributed equally across its surface.
>
> (2010, 23–24)

> how can humans become more attentive to the public activities, affects, and effects of nonhumans? What dangers do we risk if we continue to overlook the force of things? What other affinities between us and them become apparent if we construe both us and them as vibrant matter?
>
> (2010, 111)

Following Bennett's understanding of "vibrant matter", it can be argued that it is by working out the vibrancy of matter, one may be able to figure out energy-matter continuum.[6] It means that fluidity and vibrancy of matter stand in dialogue with the pervasive, invasive and diffusive flows of energy. This can be elaborated by drawing a reference to Christopher F. Jones's contention in "The Materiality of Energy": "our energy systems have become more and more materially-intensive, yet the vast majority of consumers have become progressively less cognizant of their tangible presence" (2018, 378). It points to the fact that material structures have significant bearings on the shipment of energy and also play instrumental role in shaping "possibilities for how human societies reorganize themselves around these new pathways" (2018, 387).

At this point, one may stop and think: How does energy "matter" in reality? Does the "mattering" of energy render it *energy ensemble*? Here, one may take resort to *energy thinking* which can help one explain in what ways energy "matters" in reality. *Energy thinking* encourages one to tinker with unconventional and radical thoughts so as to unravel complex operations of energy through material bodies. Whereas *energy linkages* inform one how "energy flow" forges connections between material entities, *energy diagrams* lead one to interrogate the imposition of "infrastructures" on the fluid (trans)figuration of energy. In short, transformative potentials of the geology of matter(ing) stand wedded to the fluid onto-epistemology of energy. It leads one to reach the point that neither the vibrancy of matter nor the transductive qualities of energy can be territorialized and stratified.

Mapping the Micropolitics of Energy

Micropolitics of energy[7] can hardly be mapped without taking recourse to the representation of energy in different epistemological l traditions including Indic and non-Indic traditions. For example, in *Aiteraya Brahman*, it is said: "sarvadevatyo agni" (qtd. in Kumar 2016, 296). It suggests that the god Agni happens to be a guardian of all gods and goddesses, and one has to satisfy Agni to reach other gods and goddesses. Here, Agni stands for "energy" which constitutes the Earth itself. Religious worshipping of the god Agni veraciously results in the celebration of energy which stands pervasive through all material bodies. In a way, it is by worshipping energy, one can connect the planetary world to the cosmic world. Besides this, many references to energy in different Vedas imply that people of ancient times were quite conscious of the "energy linkages" and how the natural flows of energy speak of the "interconnectedness" among geo-materialities. One may take the example of Indic notion of "Shaktism" which speaks of the "intensive" and "extensive" becomings of "shakti" through the living and the dead. Prof. Anway Mukhopadhayay has elaborately discussed Indic conception of "shakti" which, to an extent, corresponds to the Western notion of "energy" in "The Shakti Pithas: The Active Corpse, Immanent Shakti and the Sacred Geography of Shaktism": "Devi-as-corpse is a corpse endowed with energy" (2018, 72). Mukhopadhayay tries to point out that the Indic notion of "shakti" pervades across dead bodies and has a plane of immanence for its functioning. Natural agency of "shakti" has to be counted to figure out the fact that the Indic conception of "shakti" does not only speak of "extensive" mattering but also suggests an "intensive" mattering. Besides, one may be reminded of how aesthetic dimensions of energy are held as quite important in the realization of "heroic" rasa. Prof. Ramaranjan Mukhopadhayay has succinctly explained in *Rasa Samīkṣā* that being one of the nine permanent feelings, "energy" gets transformed into "heroic" rasa when determinants, consequents and transitory feelings work together (2001, 56).

Posited against Indic understanding of energy, one may be reminded of how energy is figured out in Western thinking traditions. The word "energy" is derived from the Greek word "energeia" meaning "enworkment", "putting-at-work" and "activation" (Marder 2017, 13). Here, the prefix "en-" means "in" and "ergon" means "work", so, it is clear that energy has something to do with "work". Although Aristotle has discussed "energy" in *Physics* and in *Metaphysics*, he has not ingeniously attempted to translate it into English and leaves it up to readers to make philosophical speculations relating to the ontology of energy. Interestingly, Aristotle has not attempted to equate "energy" with "potentiality" or *"dunamis"*, for the latter is steeped in the notion of "activation". Unlike "potentiality", energy bears epistemic leanings to "actuality" (Marder 2017, 16). In a sense, whereas "activation" is suggestive of the process of being in an "activity", "actualization" entails energy's being in an "actuality". "Potentiality" means "power" or "capacity" that is required to do "work". But, Aristotle's understanding of "energy" exists beyond the "vicissitudes of dunamis" (Marder 2017, 18). The notion of "enworkment" also suggests that energy being an agent of "actuality" engages one in "enworkment". Here one may find two possibilities: Either "enworkment" implies that energy sets one to work or it is the upshot of a product and draws one into "actuality". Therefore, an understanding of the being of "energy" is contingent upon energy's transduction from one material form to another. In a way, "energy" is something that spells out one's "-at-work(ness)".

At this point, one may stop and ponder over the differences between *energy infrastructure* and *energy diagram*—whereas the former refers to a series of networked mechanisms and structured organizations by means of which energy gets transducted and transported, the latter does not speak for any epistemic codification; rather, it allows energy to assume diverse forms thereby helping one consider a plethora of possibilities. *Energy diagram* thus is implicative of a fluid figuration of energy and helps one bear out the seamless and processual transference of energy from one form to another. It is true that energy is stored in one form and hence has certain form-related structure. In the next moment, the natural conversion of energy from one state to another makes it appear "problematic" or what can be understood as an *energy paradox*.[8] It is by employing "process philosophy",[9] one may contend that fluxional movements of energy are capable of generating "power" required for doing an activity. As the functional variety of energy makes it oscillate between deterritorialization and reterritorialization, energy opts for a radical alterity or what can be understood as an *energy heterodox*[10] in terms of movement and process. Action Network Theory (ANT)[11] provides the idea that the existence of everything in both the human and non-human world is within a fluid network of relationality and is, therefore, teasingly problematic. Bruno Latour, one of the pioneers of ANT, reflects that ANT can only be used to "describe" events which are always in movement either "intensively" or "extensively". Since the ontology of energy speaks of its "differential intensity", ANT could be exploited to lay out the functional positionality and positional functionality of energy in a network of relationality. This processual transition of energy reminds one of a Deleuzo-Guattarian formulation called "rhizomatics". Just like a rhizome unfolds itself in an aberrant way, energy flow is also marked by aberration and precarity, heterogeneity and transmissibility, connectivity and deterritorialization.

In *Energy Humanities: Current States and Future Directions*, Mišík and Kujundžić, while mapping various objectives of energy humanities, reflect:

"energy humanities seeks to identify the social, cultural, and political changes necessary to facilitate a full-scale energy transition, anticipate their consequences, and imagine possible scenarios for a future" (2021,10). This reflection points to the fact that "energy flow" seeks to impact not only functional and organizational aspects of a society but also political, economic and cultural becomings of a society. It leads one to arrive at the rhizomatic or diagrammatic progression of energy which entails "lines of flight"[12] (Deleuze and Guattari 1987, 11), that ultimately leads one to figure out radical and revolutionary "potentials" of energy. At this point, one may stop and think: What is *energy linkage*? How can *energy linkages* be understood in epistemic terms? Are *energy linkages* the end points and beginning points as far as energy flow is concerned? Why is it so important to take *energy linkages* into account? Critical responses to these unsettling questions posed in the subsequent paragraphs can help one find the worth of *energy linkages* in comprehending the matrix of "energopolitics".

Energy linkages can be epistemized as *fluid nodes*—the construction of which depends on random flows of energy. Just like in a rhizome, "nodes" either produce shoots and roots or shoots and roots end up in "nodes", *energy linkages* bear epistemic congruity with the formation of "nodes". Put in other words, like "nodes" in a rhizome, *energy linkages* neither take particular structural patterns nor let one put it in structural patterns; rather, they are characterized by porosity and precarity. One may further work out the logic of contiguity to account for the formation and functionality of *energy linkages*. Jean Luc-Nancy has come up with the idea of "being with" in *Being Singular Plural* "if Being is being-with, then it is, in its being-with, the 'with' that constitutes Being; the with is not simply an addition. This operates in the same way as a collective" (2000, 30). Samuel Weber moves a step forward and moots in *Singularity: Politics and Poetics*: "the singular event—the event as singular … is … a process that seems to be its direct contradiction: a process of repetition. But it is a repetition that is composed not just of similarity, but of irreducible difference" (2021, 2). Following two distinct critical interventions into the contiguity of "Being", it can reasonably be contended that *energy linkages* are premised on the logic of contiguity and thus are replete with potentialities and possibilities. In a nutshell, *energy linkages* are *transterritorial* in nature and therefore, refuse categorical mappings. *Energy linkages* dwell in the *energosphere*[13] which is a deterritorialized geo-actuality.

Keeping the idea of *energy linkages* in mind, one may put forward that biopolitics nowadays takes a back seat, allowing "energopolitics" to rise to the occasion inasmuch as energy as micropolitical engagements with human and non-human entities across the planet.

Whereas Foucault's formulation of biopolitics derives "its knowledge from, and its power's field of intervention in terms of, the birth rate, the mortality rate, various biological disabilities, and the effects of the environment" (*"Society Must Be Defended"*, Foucault 2003, 269), Dominic Boyer takes this further and configures "energopolitics" as an epistemic derivation of biopolitics. Boyer reflects that "energopolitics" is premised on the dual analytics of energy and politics, and thus it can reasonably be resorted to "unmake and remake" (2019, xiv) the domain of anthropocentric political theory. Imre Szeman contends that "energopolitics" could at once be taken into account as "a fuller account of biopolitics" (2014, 460) and at times be held as one that "addresses problems and limits" (2014, 460) in biopolitical governance propounded by Foucault. Szeman has also mapped "energopolitics" in terms of "energopower" in "Conclusion: On Energopolitics": "Energopower is not an alternative because it replaces biopower, but because it insists on the necessity of examining the essential function of energy in 'the organization and dynamics of political forces across different scales'" (2014, 456). Szeman means to say that contemporary society undergoes an energy crisis of "a different kind and on a different scale" (2014, 456) and recognition of energy both in governing the (form)ation and trans(form)ation of subjects and in configuring the organization and reorganization of cultural, political, and economical discourses is a timely necessity. Boyer, too, agrees that the "ethics and epistemics" (Boyer 2019, 15) of "energopolitics" cannot be deciphered if one does not take into account "energopower". At this point, one may take a stop and mull over the differences between "energopower" and "biopower". Whereas "biopower" is chiefly exercised in governing the "bios" thereby facilitating biopolitics to assist the functioning of a government, "energopower" primarily centres on the transductive and transformative, invasive and diffusive potentials of energy, and it happens to be the operative impetus of "energopolitics" in reality. Boyer contends that "energopower" provides a "way to join together discussion of emergent postneoliberal political potentialities with the energic forms of "revolutionary infrastructure" that will necessarily enable them" (2019, 16). Boyer's argument leads one to arrive at the conclusion that the neoliberal government nowadays seeks to employ exploitative mechanisms aiming to govern the expanse of energy that lies beyond the earthly or planetary expanse of "bios". Micropolitical flows of energy also suggest that the advent of energopolitical governmentality is arguably premised on the strands of "energopower" which caters necessary impetus to a neoliberal government for exploiting the energy-matter continuum.

Epistemic and governmental switch to the "energopolitics" from the prevailing biopolitics eventually entails free and unrestrained anthropocentric exploitations of planetary resources which include ecological and marine resources among others. It has become easier for the neoliberal government to regulate the tension between energy and economics insomuch as "energopolitics" facilitates neoliberal government to capitalize the concentration of energy stored in various geological forms, at different locales. This argument could be elaborated by referring to *Energy at the Crossroads: Global Perspectives and Uncertainties* where Vaclav Smil points out the profound importance of "energy-economy link" in the *energo-sphere*:

Relationship between the quality of life and energy use is perhaps even more complex than is the energy–economy link ... there are no preordained or fixed levels of energy use to produce such effects. Irrational and ostentatious overconsumption will waste a great deal of energy without enhancing the quality of life while purposeful and determined public policies may bring fairly large rewards at surprisingly low energy cost. Seaborne transport of crude oil is a leading source of polluting the ocean waters, particularly of the coastal areas.

(2003, 64)

In short, unlike biopolitics, "energopolitics" in actuality enhances governmental exploitations of earthly resources other than "bios" by means of empowering a neoliberal government to exceed the bounds of biopolitical governmentality and to go to the extent of more nuanced and subtle micropolitics of energy which transversally cut across spatio-temporalities and post(g) localities.

Interrogating Dominic Boyer's *Energopolitics*

Both energopower and biopower . . . are entering into a pivotal transitional phase. ("Energopower: An Introduction" 2014, Boyer 309)

Boyer's formulation of "energopolitics" although promises to be fruitful for the smooth governing of planetary species along with influencing the becomings of the Earth, it ultimately proves to be unproductive of anything "new". It actually corroborates Foucauldian formulation of "biopolitics" in subtle ways and has virtually no effect on the ongoing practice of governmentality. In short, Boyer's "energopolitics" is fraught with conceptual lacuna and cannot do anything good for the wellbeing of the planetary entities on the Earth. In order to explore the limitations of Boyer's "energopolitics", one may take recourse to the following questions: Where does the inoperativity of "energopolitics" lie? Why does "energopolitics" fail to comply with the production of the "new"? Does "energopolitics" have bearings on the depletion and disappearance of marine resources in particular? Although Boyer aims to delve deep into the micropolitics of energy thereby surpassing the bounds of "bios", his critical coinage fails to get rid of the lasting influences of Foucauldian biopolitics and thus it turns out to be an epistemic exercise in redundancy. Although the conceptualization of energopower by Boyer seems to be effective in laying down a ground for "energopolitics" to operate with full capacity and efficacy, "energopolitics" ultimately ends up being a veritable state politics which is put into practice to regulate the flows of "energy capital" through earthly entities. Moreover, Boyer's "energopolitics" is restricted within the boundaries of "aeolian politics" and does not directly interact with folding and unfolding of energy through the processual transformations of different earthly "matters". Although "energopolitics" seems to turn away from post-anthropocentricism by means of embracing the profound importance of the micropolitical engage-ments of energy with earthly matters, biopolitical moorings of "energopoli-tics" prevent it from being wedded to the production of the "new", thereby forcing it to comply with the hegemonic structures of state politics.

The inoperativity of "energopolitics" primarily lies in its incapacity to break free from the discursive limitations of biopolitics and to provide a sort of radical alterity which is instrumental in interrogating governmental exploi-tations of earthly matters as a whole. *Energy linkages* and *energy thinking*, in particular, remind us of how governmental attempts are frequently made to force"infrastructures" on the fluid dissemination of energy through earthly matters and it becomes obvious in the example of human "infrastructures" for extracting maximum "energy capitals" being built in an ocean. In a way, "energopolitics" insists on the containment of "energy capitals" so as to help the government keep energy-economy tension under check. Besides, one may argue that Boyer's "energopolitics" cannot provide a way-out for planetary beings who are often subjected to crass commoditization of energy. The inop-erativity of "energopolitic" in maintaining planetary equilibrium lies in the fact that it actually upholds capitalistic manipulation of the flows of energy to satisfy human needs and desires.

Boyer's "energopolitics" in actuality fails to provide ecologically sound frameworks for political governance, and it conforms to the operative strands of global capitalism which work toward maintaining "energy pov-erty" ("Fossilized Liberation: Energy, Freedom, and the "Development of the Productive Forces", Huber 2018, 516). Fluid and elusive dimensions of energy attract global capitalists who nowadays turn to oceanic resources—global marketing of which lays stronger grounds for the dissemination of energy. Rather, freeing "energy flow" from any human channelization is required to set the economic system of a society in the right direction. In order to substantiate this observation, one may be reminded of Allan Stoekl who pertinently reflects in "Marxism, Materialism, and the Critique of Energy": "[the] transformation of the economic system will result in a freeing of ener-gies and minds and a more egalitarian distribution of wealth" (2018, 16). It is true that illicit supply of energy proliferates illegal marketing which is quite vicious to the marine and coastal ecologies at large.

It can be contended that the strands of "energopolitics" do not just only stand confined in "aeolian politics" but also seem to be at odds with the nuanced intersections between marine and coastal ecologies. Energopolitical strands look quite ruinous to the abundant oceanic resources in the sense that these strands help global capitalists to carry out ruinous strikes at the smooth interactions between marine and coastal ecosystems. Densely inter-connected marine and coastal ecosystems that are grounded in the *energy linkages* stand vulnerable to unsolicited human interventions into the oceans. Governmental appropriations of *energy linkages* backed up by "energopo-litics" need to be questioned to reinforce the profound importance of the geology of matter(ing). Although Boyer's critical formulation of "energopolitics" stands riddled with conceptual gaps and epistemological limitations, the relevance of *energy thinking* can hardly be denied in this context. Instead of complying with the insular configuration of "energopolitics" propounded by Boyer, one needs to critically engage with the micropolitics of energy supported by the geology of matter(ing) to free "energy flows" from any sort of territorialization and stratification. In doing so, one may build up a critical theory of energy to interrogate the crippled paradigms of "energopolitics". At this point, one may be reminded of Imre Szeman's influential article titled "Towards a Critical Theory of Energy" where Szeman critically reflects:

> energy matters—can be explored in any number of ways … energy humanities articulates a demand that we reimagine the vocabularies, methodologies, and presumed objects of study of the disciplines as they currently exist, largely because they were constituted in the absence of an essential component of human experience: energy.
>
> (2021, 24)

Following Szeman's critical reflection, one may plausibly put forward that "energopolitics" eventually gets turned out as an exclusive and inflexible epistemological standpoint that presupposes the operation of energy within a stratified territory and thus, the matter(ing) of energy needs to be taken into consideration to de-structure "energy flow" across political, cultural, geological, social and economical aspects, among others, with the help of the inclusive and deterritorial epistemological frameworks provided by restructured Blue Humanities.

Notes

1 In "Fifty Key Terms in the Works of Gilbert Simondon", Jean-Hugues Barthélémy defines "allagmatics" as "the theory of operations" (2012, 204). It refers to a particular kind of operation that "makes a structure appear" (2012, 204) or "modifies" (2012, 204) a structure. Unlike other scientific theories of operation, "allagmatics" brings out "intensive" movement of a structure.

2 Dominic Boyer happens to be the critical theorist who has elaborated on different aspects of "energopolitics"'in his ground-breaking work *Energopolitics: Wind and Power in the Anthropocene*. In this book, he engages himself in exploring the importance of energy in a time when the Anthropocene strives to find alternative sources of natural energies to help sustain human civilization for a longer period of time.

3 Ghosh holds "trans-plastic-habit" as a sort of entanglement that "forms" (2021, 120) and has the ability to "receive and give form" (2021, 120). Trans-plastic-habit refers to a distinctive form(ability) which "gives, sculpts, gestaltizes, imbibes, fractalizes, explodes, experiments and maintains form" (2021, 120).

4 It basically refers to the continual flow of energy through different geo-materialities. Energy is something that always stands linked with human and non-human entities across the world.

5 *Energy ensemble* can simply be understood as a collective of energies. Several energies seek to overlap at the point of an ensemble.

6 Whereas in "A Philosophy of Energy", Stanley Jacobs understands energy as a part of a "continuum" (1989, 95) and discusses it in terms of "stillness", "activity" and "accumulation" (1989, 95), in "Towards a Philosophy of Energy", Robert-Jan Geerts et al. work out a philosophy of energy in terms of making inquiries into "the natural phenomenon of energy", "the functioning of energy in society" and "philosophy of technology" (2014, 108). Unlike Western tradition of thought, in the realm of Indian philosophical traditions, energy is held tantamount to "Śakti" which "means "power"; in Hindu philosophy and theology Śakti is understood to be the active dimension of the godhead, the divine power that underlies the godhead's ability to create the world and to display itself" (*Hindu Goddesses: Visions of the Divine Feminine in the Hindu Religious Tradition*, Kinsley 1988, 133). Śaktism insists that Śakti (power) has to be taken into account as a supreme feminine power ("Devi" in Tantric tradition) that is responsible for all earthly and spiritual activities. For instance, one may find detailed analysis of "Shaktism" in "The Shakti *Pithas:* The Active Corpse, the Immanent Shakti and the Sacred Geography of Shaktism" by Anway Mukhopadhyay.

7 Ontology of energy helps one grasp the distinctive "being" of energy. Here, one may mark that energy possesses "differential being" which makes it fluid, dynamic and processual in nature. In *"Energeia, Entelecheia,* and the Completeness of Change", Mark Sentesy elaborately discusses both "energeia" (2020, 66) and "entelecheia" (2020, 68) in relation to the ontology of change and argues that an understanding of the ontology of energy cannot be complete if "energeia" (2020, 66) and "entelecheia" (2020, 68) are left out of consideration.

8 *Energy paradox* is configured to capture the problematic dimensions of energy. Paradoxical presence of energy in a material form for the time being, coupled with its natural intensive flow, makes it a contentious phenomenon.

9 Alfred North Whitehead happens to be the pioneer of Process Philosophy which seeks to spell out how reality gets unfolded in the manner and mechanism of a "process" (1978, 208). Later on, several critical thinkers including Thomas Nail and Gilles Deleuze, among others, contributed to this field of study.

10 *Energy heterodox* can be understood as a kind of radical alterity which energy runs after. The natural tendency of energy to opt for a heterodoxical position renders it revolutionary and rebellious against territorial mapping. It also suggests that any attempt to make epistemic incarceration to energy ultimately falls flat because of this phenomenon called *energy heterodox*.

11 Action Network Theory (ANT) basically refers to the fluid relationality between agents in a network. In *Reassembling the Social: An Introduction to Action Network Theory*, Bruno Latour has extensively dealt with the distinct "political epistemology" (2005, 249) of ANT.

12 "Lines of flight", in the view of Deleuze and Guattari in *A Thousand Plateaus*, can be comprehended as "ruptures" (1987, 22) which condition the revolutionary becomings of a rhizome.

13 *Energosphere* bears epistemic congruity with "rhizo-sphere" in the sense that *energosphere* is devoid of strata and a "plane" which paves the way for transversal becomings of energy. It has multiple entries and exits and is pervasive in nature.

V Sri Lankan *Minor* Fiction
Earth(ing), Energy Flows and Oceanic Ecologies

Introduction: Interfacing Energy, Ocean and Earth(ing)

the Mechanosphere ... the abstract Machine of which each concrete assemblage is a multiplicity, a becoming, a segment, a vibration. And the abstract machine is the intersection of them all.

(A Thousand Plateaus: Capitalism and Schizophrenia,
Deleuze and Guattari 1987, 252)

Our increasing awareness of climate change is catalyzing new imaginaries and, by extension, new allegorical forms to address the dynamism of our planet. (Allegories of the Anthropocene DeLoughrey 2019, 1)

It is by now clear that in the context of oceans, energy plays an important role in causing different marine ecosystems to interact with each other. Interestingly, the ocean happens to be a distinctive planetary entity that allows energy flows to go in different directions, resulting in the mobilization of the ocean. In other words, it is because of the constant energy flows found in the oceanic ecological interactions, the ocean ceases to be a static entity and stands in tandem with the deterritorial flows of the Earth. Needless to say, territorialized understanding of the energy simply does not work in the context of the ocean which makes room for free flow of energy through diverse marine ecosystems that stand rooted in the geokinetic becomings of the Earth too. In order to materialize the interface among energy, ocean and earth(ing), certain South Asian fictional narratives are intentionally drawn to show how in the domain of literature, complex intersections among energy, ocean and earth(ing) get represented.

The objective of this chapter is therefore chiefly two-fold: It is by considering select Sri Lankan "minor" writings, the nuanced interactions and intersections among energy, ocean and earth(ing) are critically investigated and in doing so, it aims at offering a critical theory of energy by means of contextualizing "energy thinking" in select Sri Lankan fictional narratives. In order to carry out this objective, theoretical inputs elaborately spelt out in the previous chapters are to be frequently drawn from through this chapter's critical engagement with select fictional narratives.

DOI: 10.4324/9781032629728-6

It has now become quite clear that a "minor" writing is of profound importance, for it affords a political space for making a series of reconfigurations and realignments. Following the immense political potentials of a "minor" narrative, one may plausibly argue that a "minor" writing can stage the interactions and intersections among energy, ocean and earth(ing) and in addition to it, a "minor" writing gives room to an individual to make direct access to the differential movements of energy, ocean and the Earth. It is by means of making geophilosophical interventions into the processes of earth(ing), frayed edges of Blue Humanities could be restructured so as to make it function as a "war machine" against the dynamics of "energopolitics". In other words, a complex assemblage of relationality, heterogeneity, productivity, co-extensiveness and "compossibility" needs to be closely examined to bring out some possible ways of turning "oceanic thinking" into an effective instrument for interrogating the workings of "energopolitics" at praxis. Using the theory of "assemblage", it is to be argued in this chapter that fictional instances are intended to taken into account from select Sri Lankan narratives to corroborate critical contentions pertaining to the nuanced interplay among energy, ocean and earth(ing).

Now, one may stop and ponder over the following questions: What is an assemblage?[1] How does an assemblage help one connect three diverse entities like energy, ocean and the Earth? An assemblage is evocative of a multiplicity which comprises various intensities that hold a multiplicity back from being codified. Deleuze and Guattari conceptualize the "abstract machine" to lay out how an assemblage functions as a multiplicity which speaks of "convergence" and "connectivity" in the "Plane of Immanence". Thus, they cogently reflect on the "Mechanosphere" which is held responsible for the eradication of strata and codes, delimitation of territorial boundary and inclusivity of the abstract machine. They reflect that consideration of the "'Mechanosphere" can help one figure out how the operations of different "abstract machines" are "convergent" and "intertwined" (1987, 514). This contention seems to be helpful for one to understand how the "intertwined" functionings of energy, ocean and earth(ing) comply with the logics of "convergence" and "connectivity" in the "Plane of Consistency". Hunter Dukes has rightly contended in "Assembling the Mechanosphere: Monod, Althusser, Deleuze and Guattari": "The mechanosphere opens the earth to the possibility of recombinatory destratification: the radical (and potentially dangerous) re-assemblage of matter into novel collectives" (2016, 522). Following Dukes's intervention, it can be argued that the "Mechanosphere" is able to do away with rigid segmentarities thereby allowing the Earth to function as a body without organs (BwO).

Romesh Gunesekera's Reef

Thus, for Gunesekera, Sri Lanka, despite all its physical beauty is not a "paradise." ("Landscape in Romesh Gunesekera's *Reef, The Sandglass*, and *Heaven's Edge*" 2017, Daimari 63)

Romesh Gunesekera carves out a singular position in the gamut of Sri Lankan literature by producing a number of remarkable fictional and non-fictional narratives including *Reef* (1994), *The Sandglass* (1998), *Heaven's Edge* (2002), *The Match* (2006), *The Prisoner of Paradise* (2012), *Noon Tide Toll* (2013) and *Suncatcher* (2019). *Reef* relates the story of Mister Ranjan Salgado who happens to be a marine biologist by profession and has a considerable amount of interest in Sri Lankan cuisine. Mr. Salgado appoints Triton as his servant in the house. Triton used to cook different kinds of Sri Lankan dishes for him. The plot of this fiction veers around how Mr. Salgado relishes different kinds of Sri Lankan dishes prepared for him by Triton, his reflections on the various ways by means of which the Earth gets polluted, his frequent association with Miss Nili, his deep-seated longing for Sri Lanka through the avenues of memory, his observations on the massive Sri Lankan Civil War, destruction and disappearance of reefs under the sea and so on. Owing to the multidimensionality of *Reef,* scholars have approached it from various critical standpoints. For example, *Reef* is often read as a fictional specimen of *Bildungsroman,* a novel of formation, for it reflects on the growth of Mister Ranjan Salgado as a marine biologist and an entrepreneur who wishes to tap into the free flow of capital in the domain of the global tourism industry. Sometimes, *Reef* is studied from the perspectives of Diaspora Studies in that Mr. Salgado tries to connect himself with Sri Lanka in terms of memory, food and belonging when he stays abroad. *Reef* is frequently referred to as a distinctive Sri Lankan narrative that uncovers the gradual decimation and disappearance of reefs under the sea, due to inordinate human interventions. When *Reef* is examined from green criminological viewpoints, it is argued that Mr. Salgado often acts like a green criminal who exploits green resources for lucrative gain. Unlike these critical interventions, this present study seeks to lay bare how *Reef* stands possessed by the configuration of a "minor" writing in general, and particularly how it fictionalizes "convergence" and "connectivity" among energy, ocean and earth(ing) in the "Plane of Immanence".

Reef actually turns out to be a fictional rejoinder to the rapid depletion and disappearance of coral reefs, for it allows readers to be introduced to the ways certain human beings being fuelled by global tourism and capitalism engage themselves in putting the sea at risk. This fiction opens up with a pointed epigraph—"Of his bones are coral made" (*The Tempest*, Vaughan and Vaughan, 2011, 200) which, originally taken from William Shakespeare's *The Tempest*, refers to the immanent "connection" and "convergence" between the human world and non-human worlds under the sea. In other words, this excerpt underlines the extensive processual (en)folding of a coral reef in conformity with the passage of time. The formation of coral reefs under the ocean generally takes a lot of time and this slow and natural process stands quite vulnerable to the onslaughts of corrosion and erosion. Besides, a coral reef happens to be one of the most frail and irretrievable marine resources—the depletion of which gets worse due to direct human contact. Continual mutation of coral reefs coupled with its speedy depletion in the age of neoliberal globalization stands as a serious cause of concern for blue humanists in particular, for massive changes in the marine ecological system inevitably result in "a sea- change" (*The Tempest*, Vaughan and Vaughan, 2011, 200) in the terrestrial ecology. At the inception of the fiction, a pointed reference to this Shakespearean expression is supposed to suggest that readers are going to be introduced to the issue of how marine resources, including reefs, are at jeopardy and coastal ecology, in particular, plays an instrumental role in articulating constant human threats to marine ecologies. The introductory chapter is also very interesting not only because it sets the tone of the fiction but also because it orients readers to come to terms with linguistic singularities characterized by deterritorialization, which this fiction seeks to set forth. Conversations in the introductory chapter—titled "The Breach"—remind readers of how this fictional narrative turns away from any definite linguistic patterning and inclines to be governed by linguistic decentralization, transgressivity and fluid referentiality. This is reflected when Triton, the servant of Mr. Salgado, strikes up a conversation with an unknown immigrant from Sri Lanka and uses a kind of linguistic expression which is not common and, more importantly, does not conform to any definite linguistic pattern of expression: "'you in this country a long, long time then?'" (Gunesekera 1994, 2). In a way, readers are linguistically prepared to confront a number of "breaches" metaphorically as the tale of Mr. Salgado begins to unfold.

The first chapter is titled "Kolla", thereby reminding readers of "kola kanda", a Sri Lankan food that is generally served at breakfast time. As the story progresses in the opening chapter, readers are gradually informed about Mr. Salgado's fascination with Sri Lankan cuisine in general and different Sri Lankan recipes which interest him. It appears that references to the various food items are made just to underscore Mr. Salgado's intense fondness for Sri Lankan cuisine but a deeper understanding of the opening chapter helps one figure out how Gunesekera includes typical Sri Lankan food jargon to authenticate the fictional narration of Mr. Salgado's inclination to Sri Lankan cuisine, in particular: "*seeni-sambol*" (1994, 8), "*pol-sambol*" (1994, 15), and so on and so forth. The use of typical Sri Lankan food jargon couple with linguistic transgressivity makes this fictional narrative conform to the configuration of a "minor" writing.

Mr. Salgado happens to be a "self-educated" (1994, 24) person who firmly believes in the deterritorialization of knowledge and thus studies "mosquitoes, swamps, sea corals and the whole bloated universe ... legions under the sea, the transformation of water into rock—the cycle of light, planton, coral and limestone—the yield of beach to ocean" (1994, 24). In spite of being a marine biologist, Mr. Salgado finds considerable interests in the process of earth(ing) conditioned by the flows of energy in general, and particularly the dwindling conditions of coastal ecology—a liaison between marine ecology and terrestrial ecology. Being driven by the "immanence" of knowledge, he engages himself in exploring interdisciplinary perspectives to examine the differential intensities of the ocean. This gets reflected when he tries to connect chemical energy and a "vicious strain of mosquito" (1994, 34) with the differential progression of the Earth: "The mosquito is a much-neglected beast. If we fail to study it and simply rely on DDT, we do so at our peril" (1994, 34). This suggests that Mr. Salgado is quite aware of "energy linkages" and how they strikes up a balance between human and non-human worlds on the Earth.

Disruptions in "energy linkages" caused by the uncontrolled spreading of Dichlorodiphenyltrichloroethane (DDT) for the containment of the virulent strain of mosquito ultimately result in the transformation of the Earth.

Gunesekera has ingeniously woven political references with the development of the plot thereby suggesting that every occurence of "micropolitics" in this fictional narrative stands in tune with the macropolitical alterations that were happening in Sri Lanka. For example, one may find in this fictional narrative that in the garb of research, Gunesekera makes Mr. Salgado ddivulge harrowing tales relating to the coral reefs which are dwindling and disappearing very rapidly and more shockingly, the government, in spite of being aware of the depletion of coral reefs, keeps mum. It is by the help of "micropolitics" that Mister Salgado is able to stand in non-conformity with the prevailing governmental "macropolitics" ratified by the one-sided and inequitable workings of "*men*: the Foundation" (1994, 48). This "minor" fictional narrative itself reckons on the revolutionary potentials of "micropolitics" in inspiring intellectual questions on the accountability of a foundation like "men" and the government, which, instead of revealing the harsh truth concerning the decimation of coral reefs, work hard to conceal it time and again.

Turning against the grain, Mr. Salgado pertinently observes:

> Anything! Bombing, mining, netting … you see, this polyp is very delicate. It has survived aeons, but even a small change in the immediate environment—even *su* if you pee on the reef—could kill it. Then the whole thing will go. And if the structure is destroyed, the sea will rush in. The sand will go. The beach will disappear.
>
> (1994, 48)

This fictional excerpt points out how a subtle change in the marine environment (including any minor human intervention by means of peeing) can result in the decimation and disappearance of polyps. It also reflects on how varied exercises of "energopolitics" in terms of "bombing", "mining" and "netting" condition the jeopardization of endangered sea species like coral reefs, polyps and so on. Mr. Salgado has rightly argued that it is because of the government's sheer indulgence in the "energopolitics" that no serious step is taken to contain human interventions in the oceans. Interestingly, Mr. Salgado points out how the disappearance of engendered sea species expedites the corrosion of sea beaches, the rise of sea level and decimation of the coastal ecological system. He understands it quite well because "energopolitics" is severely ruinous to the wellbeing of the oceans in general and has a terrible impact on the coastal ecology. Thus, it seems very important for Mr. Salgado to safeguard coastal ecological frameworks by means of considering the nuanced intersections among the ocean, geokinesis and immanent trajectories of energy.

One may be reminded of another interesting passage in the text where the free flow of capital energy instrumentalized by the boom of global tourism is lambasted on the ground that it has the potential to usurp existing national-ist economic frameworks in Sri Lanka and to put the young generations at jeopardy. Posited in other words, the indigenous Sri Lankan economy is, to a large extent, contingent upon the revenues from the local tourism industry but the invasion of capital energy through the networks of the global tourism industry proves vicious to the sustainable growth and development of marine resources found in Sri Lankan coastal regions. Global capital energy turns the Sri Lankan young generation into "servants" (1994, 111) who ultimately become victims of the politics of (g)locality. Mister Salgado minutely reflects:

> These people all think tourists will be our salvation. All they see is pockets full of foreign money. Coming by the plane-load ... They will ruin us. You know, brother, our country really needs to be cleansed, radically. There is no alternative. *We have to destroy in order to create*. Understand?
>
> (1994, 111)

With the backdrop of ongoing the Sri Lankan Civil War, Mr. Salgado pre-cisely points at the "radical" reversal that might be of profound importance in reconfiguring society at large. He insists that an engagement with "micropo-litics" is required to hold contemporary Sri Lankan society back from being subjected to territories and codes. It is clear to all that Gunesekera subtly makes use of the politico-economic backdrop to legitimize the concerns of Mr. Salgado for economy-energy duality.

Connections between capital energy and endangered marine resources are articulated when the crab-seller informs Ms. Nili that someone has "caught a dolphin" (1994, 118) and "will kill it quickly" (1994, 118) for it can help him earn "Very good money" (1994, 118). It shows that endangered marine species are commoditized in the local market, which has international con-nections owing to the exposure of rich Sri Lankan coastal lines. In a way, being driven by the forces of "energopolitics", under the nose of govern-mental authority, illicit blue trafficking of endangered species like dolphin and reefs, among others, takes place. On the one hand, the global tourism industry exposes the rich concentration of marine resources at Sri Lankan coastal regions, and on the other hand, it invites blue traffickers to engage impoverished local youths in the (g)local exportation of marine resources: "Outside a man was filling an unmarked van with baskets of dead fish. Small pieces of bleached white coral marked the municipal parking lot" (1994, 118). It suggests how "blue archives" under the ocean are time and again invaded by means of exploiting "energy linkages" to jettison the huge importance of "blue memory" and "blue trauma" in reevaluating the damages already done to the ocean.

This fictional narrative tries to be vocal against "Conspicuous consumption" (1994, 135) referring to the (g)locally emergent consumerist culture supported by the spree of neoliberal economy. Moreover, it seeks to speak up against the vicious impinges of neoliberal economic frameworks by means of taking "collective assemblage of enunciation" into account. This is clearly reflected when Ranjan Salgado goes to act out his radical thoughts not for the sake of himself but for the sake of the Earth. While staying in London, Triton notices that the ocean surrounding his cottage has been polluted, resulting in the flowing of "dead sea urchins" (1994, 171) around the beach. Much worse, he observes that "[the] sky would redden, the earth redden, the sea redden" (1994, 172). This fictional reference attests to the immanence of the Earth that lets different intensities end up in a "collective assemblage of enunciation". Nuanced connection among the sky, sea and the Earth reminds one of a BwO which stands divested of strata and territory. A rhetorical question that Triton asks indicates the pervasive trajectories of "oceanic connectivity" across the Earth: "Do all the oceans flow one into the other? Is it the same sea here as back home?" (1994, 172). Thus, any damage to a marine ecosystem inevitably results in changes in the adjacent ecologies of the oceans across the world.

Gunesekera's deep concerns for the dwindling conditions of the coastal regions get intensified when a sensitive and ethically upright marine biologist like Ranjan Salgado, shockingly and disgracefully, gives in to the snares of global consumerist culture by means of expressing his desire to build a "marine park":

> I used to think that in a month or two, the next year, I would have a chance to turn the whole bay into a sanctuary. A Marine park. I used to plan it in my head: how I'd build a jetty, a safe marina for little blue glass-bottomed boats, some outriggers with red sails, and then a sort of floating restaurant at one end. I thought of it like a ring, a circular platform with the sea in the middle.
>
> (1994, 177)

This fictional excerpt reflects how capital energy forces a sane marine biologist like Mr. Salgado to yield to vicious and injurious entrepreneurship. It ironically suggests that although Mr. Salgado was once vocal against the practices of "energopolitics", he gradually turns out to be a part of it. Taking recourse to the interactions among energy, ocean and earth(ing), one may cogently put forward that a combination of "energy thinking", oceanic connectivity and geophilosophy can be employed to interrogate the forces of "energopolitics" at play. Following this formula, Mr. Salgado's contemplative involvement in the construction of a "marine park" could manifest in such a way that "energy thinking" has to be employed to safeguard "2coastal ecologies" from imminent threats of destruction, plastic pollution and slow erosion so that the "blue archives" could be secured and the process of earth(ing) does not get interrupted.

Gunesekera's "minor" narrative, that is, *Reef* thus offers relevant insights to readers to reconfigure the epistemic limits of Blue Humanities and subsequently encourages them to use the cited insights as veritable "war machine" against the marauding march of "energopolitics".

Shyam Selvadurai's *Swimming in the Monsoon Sea*

Shyam Selvadurai, one of the eminent literary stalwarts in the field of Sri Lankan literature, began his literary career with the publication of *Funny Boy* (1994), which was followed by several publications including *Cinnamon Gardens* (1998), *Swimming in the Monsoon Sea* (2005), *The Hungry Ghosts* (2013) and others. Whereas *Funny Boy* deals with the problems of adolescence experienced by Arjie Chevaratnam, *Cinnamon Gardens* takes readers through colonial days, political unrest, religious upheaval and the despair of married gay men and elite culture—all of which ended up in the construction of modern Sri Lanka. Unlike these two literary feats, *Swimming in the Monsoon Sea* puts the spotlight on how beaches are time and again vitiated by human wastes resulting in disruptions in the coastal ecologies. Set in the 1980s, this "coming-of-age" fiction relates the upbringings of Amrith, who is being raised by Aunty Bundle and Uncle Lucky. He prefers not to remember the days with his departed mother and gradually develops an interest in Uncle Lucky's Celyon Aquariums, a company meant for the business of exotic fishes. But his birthday plans suddenly seem unpromising to him with the prediction of an unexpected monsoon. When his cousin visits from from Canada, his plans get unsettled and he starts to fall in love with him.

Selvadurai's fictional narrative has already been studied from different critical perspectives. For instance, it has been read from the perspective of Queer Studies, owing to the fictional representation of gay love. Besides this, this literary narrative is often studied as a fine specimen of a "coming-of-age" fiction because it portrays the development of Amrith from childhood to adolescence. Sometimes, the problems of parenting are studied in reference to it. Unlike these hackneyed examinations, this present study seeks to bring out the fictional representations of the intersecting trajectories among energy, ocean and earth(ing).

Swimming in the Monsoon Sea is replete with references to the intersections between energy, ocean and earth(ing). For example, at the beginning of the fiction, readers are introduced to the ruinous operations of sea waves, which repeatedly hammer at the rocks that save the beach from erosion. Here, the mighty wave energy coupled with the "monsoon sea, wild and savage ...

had eaten up the beach" (Selvadurai, 2005, 1). This fictional reference con-notes the fact that coastal ecology stages the encounters between sea waves and the waning beach in particular, allowing different kinds of energy to stand at loggerheads with each other in general. Moreover, it hints at how the sea functions as a living character in the plot and constantly interacts with both Amrith and Niresh. The interaction between sea and land finds exquisite manifestations when both Amrith and Niresh visit Kinross Beach—a popu-lar destination for tourists. Their visit to the Kinross Beach exposes how it gets contaminated and despoiled by visiting tourists who throw away different kinds of rubbish here and there on the beach. The deposition of human waste on the beach ultimately gets mixed up with the receding sea waves, thereby affecting marine species. Selvadurai points to this directly in the following line: "Amrith stooped down to collect the dirt, then put it into a bin" (2005, 156). The "minor" structure of the plot makes it easier for both Amrith and Niresh to observe the changing colour and nature of the sea: "The sea, despite its swell, was a brilliant blue, tipped with emerald green" (2005, 159). It is as if the sea meets the sky at the horizon and works as a veritable BwO. At one point, they decide to take a plunge into the sea to swim around the beach but have to return to the beach as the water starts to swell up. The constant infla-tion and deflation of the sea stand in conformity with the process of earth(ing) in the sense that it at once dismantles "energy linkages" that are functional on the beach and at times lets an "energy diagram" take shape by defining and redefining the contour of the beach.

One may be reminded of another fictional reference to substantiate how human beings look to manipulate energy flow to exploit marine resources:

> Amrith could have followed him and sat with them on the beach but he stayed in the water, going under, trying to get a glimpse of the bottom. He searched for pretty shells. Occasionally he saw one but, before he could reach out to grab it, the sand shifted and it disappeared.
>
> (2005, 208)

This reflects how human energy is intentionally directed to "grab" shells, which can be pretty expensive and rare marine resources. Although human energy is used to get hold of "shells", sea waves cause the mobilization of sands, resulting in the disappearance of the "shells" from sight. The sea makes use of its own mechanism grounded in the functional logic of earth(ing) to resist humans' attempts to possess "shells".

This fiction also exemplifies how the borderline difference between sea and beach gets coalesced when "[the] sea was moving into the beach" (2005, 232). This marine movement actually stands in compliance with earth(ing) that seeks to shape "energy linkages" by means of encroachment onto the beach. It suggests that coastal ecology is of profound importance as far as the interrogation of energopolitics is concerned. In short, remapped Blue

Humanities comes into play here in that it seeks to enfold both coastal and marine ecological issues into its epistemic consideration, thereby building up strong resistance to the free operations of "energopolitics". One may consider another example to get to the bottom of this contention. At one point in the narrative, when Amrith is cleaning himself up after taking a bath in the sea, he notices an empty hut on the beach, which is filled with "some wooden crates" and "a pile of old fishnets" (2005, 233). This scattered image of indicates that someone has left it there, knowingly or unknowingly, and it ultimately makes the coastal ecology vulnerable to despoliation and damage. This act of discarding used up materials on the beach not only puts the beach at risk but also jeopardizes the endangered marine species. It further pushes one to take into account the enormous importance of coastal ecology in deciphering the overlapping trajectories of energy, ocean and earth(ing). The following textual instance attests to the fact that the different kinds of pollution on the beach end up contaminating the salt water, thereby putting the lives of marine species at jeopardy:

> Amrith was knee-deep in water. His nose wrinkled in disgust as plastic bags and cans and other bits of garbage floated around him. Still, he had no choice but to wade through this filth to get to their road. He lost a slipper and he watched at it floated away and then sank.
>
> (2005, 236)

Crass human activities on the coastal regions leave an impact on the sea and lay bare how various pollutants turn the sea into a furious and formidable entity. Plastic happens to be an anthropogenic product which requires chemical and capital energies for its manufacturing and needs human energy for its dissemination through human and non-human spaces. In a seminal article titled "Plastic Controversy", Ranjan Ghosh maps out how "plastic time" stands in connection with Earth's deep time and renders the Earth a veritable "Plastiglomerate": "Earth's deep time, hence, is increasingly invaded by plastic time. It is here that a controversial parallelism builds between the plastic materiality and biological plasticity, where running away from plastic is always already a running into plastic" (2021, no. pag.). Taking cue from Ghosh's insight, one may cogently put forward that the deposition of plastics does not only affect the aesthetic beauty of the beach but it also ruins "convergence" and "connectivity" among different ecological patterns under the seawater. It is by foregrounding the vicious impacts of plastics on the marine world that this fiction sheds light on the nuances of coastal ecology which hold terrestrial and marine ecologies together.

What is quite interesting is that *Swimming in the Monsoon Sea* conforms to the configuration of a "minor" text and thus permits different characters including the sea not to yield to the snares of "energopolitics". It is the seamless interactions between sea and beach that conditions free exchange of thoughts between Amrith and Niresh. Colloquial and informal expressions of closeness help both Amrith and Niresh to come together: "He loved Niresh in the way a boy loves a girl, or a girl loves a boy" (2005, 234). This fiction politically raises coastal ecology to the foreground, intending to build up a critical framework to interrogate energopolitical setups. Interestingly, the deep concerns of Amrith and Niresh cease to be personal when they transcend the limits of individuality. For example, deposition of used plastic bags in the sea is not only vicious to the protagonists but it also threatens to deplete the population of rare marine species. Thus, it necessitates the assemblaging of individual concerns at the collective level so that earth(ing) does not get impeded by human activities. *Swimming in the Monsoon Sea* actually offers radical thoughts by means of which human beings can be dissuaded from vilifying beaches. For example, the self-motivated cleaning up of polluted beaches by Amrith provides an example of how beaches can be safeguarded from imminent perils. The possibility of the sea getting contaminated is reduced if beaches are kept clean and clear. It implies that exploitative human activities on the beaches need to be contained to let both the coast and the ocean perform as a BwO.

Swimming in the Monsoon Sea demonstrates how "energy thinking" plays an important role in understanding how earth(ing) pushes the sea to keep interacting with the beaches. When Amrith and Niresh swim in the sea, their human energies start to interact with marine energy.

When the sea energy engulfs plastics—a mix of chemical, monetary and human energies—it impacts the becomings of sea species. This fiction points at the "energy crossroads" which can lead one to figure out geological transformation of the Earth. In other words, "intensive" becomings of the sea is contingent upon the "extensive" "geokinesis" of the Earth. This is demonstrated when Amrith could hear the roaring sounds of the sea which seem to be more "discernible" than that of the train which passed by. The "monsoon sea" is made up of "free intensities" which keeps it moving in and around the beach. Thus, it is supposed that a reference to the "monsoon sea" is made in the title to underscore the enormous importance of the sea in actualizing the interplay among different energies.

Roma Tearne's *Mosquito*

Roma Tearne, one of the most prominent Sri Lankan fiction writers in recent times, carves out a singular niche for herself by producing a number of remarkable literary pieces. Born in Sri Lanka and brought up in Oxford, she kicked off her literary career with the publication of *Mosquito* (2007), which is situated in the context of the Sri Lankan Civil War that wreaks havoc on the growing relationship between Theo and Nulani, the two protagonists of the fiction. This publication is followed by *Bone China* (2008) which seeks to explore the family saga of Grace de Silva. It is succeeded by *Brixton Beach*

(2009) that unfolds a Sri Lankan immigrant's search for her own identity in the post-7/7 scenario. *The Swimmer* (2010) deals with different diasporic issues pertaining to the settlement and acculturation in Norfolk. *The Road to Urbino* (2012), too, examines the grim and gruesome impinges of the conflict between love and war in the Sri Lankan context.

What is interesting is that Tearne's *Mosquito* has already been subjected to different critical inquiries. For example, nuanced interactions among migrancy, identity and memory have been explored from the perspectives of Diaspora Studies. There are scholars who have also examined it as a literary exploration of the Sri Lankan Civil War understood in artistic terms.

The interface between place and identity has been examined from critical points of view to underscore how love, loss and reparation interact and intersect. But, unlike these hackneyed interventions, this present study looks into the layered interactions among energy, ocean and earth(ing) so as to work out a critical framework for interrogating the operations of "energopolitics".

Tearne's *Mosquito* starts off with a description of the sea as being "was like a mirror" (2007, 1), reflecting clashes between energies on the coastal regions. For example, coastal regions are infested with mosquitoes and thus the local authority spreads DDT to control their growth: "The Ministry of Health sprayed the coconut groves with DDT to prevent outbreaks of malaria. The metallic smell drifted and mixed heavily with the scent of frangipani and hibiscus" (2007, 5).

This act of spraying DDT results in the contamination of beaches which ultimately puts the sea at risk. Here, chemical energy coupled with human energy work together to contain the spreading of malaria but at the cost of spoiling the beauty of the beaches. Sea beaches also witness the clashes between Tamils and Singhalese, the upshots of which are quite vicious to the ontical and ontological transformation of the sea: "'Ours is a very small country. Only we care about the differences between the Singhalese and the Tamils. No one understands what this fight is about. We hardly understand ourselves anymore'" (2007, 19). This fictional reference shows how human, political and cultural energies end up in transforming the geology of beaches. Put in other words, "energy linkages" have been at work on the beaches and they pave the way for the sea to receive impinges of the Sri Lankan Civil War: "Of all the places on this island … this should be the place for fresh fish. Someone told me the army drove their jeeps on to the sands, chasing a group of men. And then they shot them. They were all young, Sir. Nobody knows what they had done" (2007, 27). Violent activities of the army "left the bodies on the beach, and the local people cleared up the mess. There is always someone prepared to clean up after them" (2007, 27–28).

It suggests how on the one hand dead bodies are unthinkingly discarded on the beaches and on the other hand how these beaches are regularly cleaned up by a few responsible and sensible locals to make it a livable and hospitable place for human and non-human species. In short, this fictional instance attests to the necessity and relevance of remapped Blue Humanities, for it stands capable of unsettling and upsetting the functions of "energopolitics".

Tearne's *Mosquito* also depicts the exquisite beauty of the coastline that stands exposed to the sea: "Groups of rocks thrust their way into the sea. Giant cacti clung to the edges of the sand. Coconut palms fringed the beach, sometimes so densely that only glimpses of sea could be seen" (2007, 84). The ebb and flow of the sea carves out the coastline in such a pretty way that it attracts people to visit the beach. Opposed to this depiction, one may refer to the bare picture of the Earth which stands exposed to thoughtless human activities: "The earth was bare and wasted, without grass, without bushes, without life" (2007, 107). Earth has its own system of operation and most importantly, it is not contingent upon human interventions. But, lifeless surface of the Earth is suggestive of how human beings have ceased to take care of the Earth which, by means of its "geokinetic" movements, ceaselessly tries to evade the processes of stratification and territorialization. In other words, the Earth lays down a ground for the sea to intervene into the geology of beaches which witnesses free play of energies: "Nothing changed. The sea still scrolled up restlessly up the beach. The catamarans remained half buried the sand. When they could, after the curfew was lifted again, Theo and the girl walked on the beach" (2007, 110).

This points to the fact that it is because of the curfew imposed by the ruling government in the state that maintenance of the beaches gets severely disrupted, which results in the deterioration of coastal ecology. Viewed from the perspectives of remapped Blue Humanities, one may cogently argue that coastal regions cannot be left unclean and dirty with non-biodegradable products inasmuch as "[the] sea is full of fish (2007, 111)" and there lies a possibility that if human wastes get into the underbelly of the sea, it shall prove disastrous to all the marine species. In addition, crackdowns of the Sri Lankan Army on the political rebels have been referred to in the fiction to suggest that clashes between different energies—political, cultural, social, military and economic, among others—impact the "geokinetic" becomings of the Earth in general:

> it was not Tamil land, it was their land. The tigers had turned their submachine guns on them, sending bullets buzzing like bees. And then afterwards the rain had washed the bodies into the river the bodies had surfaced, bloated and stinking like cattle, with stiffened limbs.
>
> (2007, 129–130)

This image of mayhem hints at how the Earth is exploited to channelize and re-channelize "energy flow" through different kinds of violent activity.

This fiction also unravels how Mr. Mendis used to take his small daughter, Nulani, to visit the reef when things were quite normal: "They were going out to the reef" (2007, 149). It suggests that Mr. Mendis exploits his human energy along with capital and social energies to intervene in the "plastic" being of the sea so as to help his daughter experience the reefs. It is unquestionably true that reefs get severely affected if humans try to get closer to rare and endangered marine species. Here, it is unbecoming of Mr. Mendis who exercises "energopolitics" to have the experiences of human exploitations of marine resources and does not think of the vulnerability of endangered marine species to human touch. This intrusive mechanism of "energopolitics" needs to be questioned by means of calling the government's willingness to take apt actions into question: "'We need Samarajeeva,' said Gerard. 'He writes eloquently. Foreigners respect him. We need him to speak out against the government. What the Chief is doing isn't working'" (2007, 175). Gerard is rightly of the view that an ethical person like Theo Samarajeeva is required to speak up against the silence of the ruling government which needs to take stringent action to contain the eruption of violence on the island and to put restraints on human interventions into the depth of the sea for recreational purposes. Here, a blue humanist would argue that endangered marine products like the reefs need to be safeguarded from the onslaught of "energopolitics" by means of tapping into the deterritorializing movements of the Earth. Enormous potentials of "energy linkages" can be effective in working out an "energy thinking" model to understand how territorialization of "energy flow" results in the deterioration of the resilience of the sea. Diagrammatic trajectories of energy need to be counted to dismantle human efforts to subjugate the sea for different purposes.

One may be reminded of another significant fictional instance to elaborate on the intrusive and exploitative networks of "energopolitics". At one point in the narrative, readers are informed of how small entrepreneurs engage themselves in building up "beach restaurants" to attract tourists from all around the world, thereby making room for free interplay between energies on the beach:

A beach restaurant was being built, and oyster-shaped swimming pool with fresh water was planned for those tourists who did not want the sea. International cuisine was all that was needed. New glass-bottomed boats began to appear and old ones were being painted over. It was many years since the coral reefs held such interest. Suddenly paradise was the new currency. The island began to rescue itself, hoping to whitewash its bloody past.

(2007, 265–266).

This excerpt clearly points to the fact that the construction of beach restaurants is certainly injurious and vicious to coastal ecology not only because it redefines the "geohistory" of beaches by means of dragging elite tourists to exploit natural settings of the beaches but also because it perturbs sea-land equilibrium in terms of marine pollution and depletion of marine species. "Intensive" transformation of the sea gets affected when products of beach restaurants get precipitated down into the sea. It ends up impacting the "extensive" alteration of the Earth as well. This excerpt also points out how marine resources like coral reefs are commoditized to get tourists in beach restaurants and the "glass-bottomed boats" make it easier for tourists to invade "blue archives" at their disposal. Unrestrained human invasions into "blue archives" leave terrible impacts on the differential progression of the sea. Viewed from the standpoint of Blue Humanities, one may tenably posit that tourists' intrusions and exploitations into sea resources in terms of varied "energopolitical" means have to be brought under control so as to put a check on "blue trafficking".

The "minor" setting of the fiction is worth considering as it allows the writer to include elements of politicality and radicality in terms of critiquing human and governmental practices. Uses of local and informal expressions help the writer to experiment with radical thoughts so as to expose disruptive and disjunctive interactions among energy, sea and earth(ing). Besides, fictional references to the enormous political event like Sri Lankan Civil War and its connectivity with the becoming of coastal ecology make this text conform to "minor" configuration. Immanent micropolitics of the Sri Lankan Civil War push Theo and Nulani to transcend the bounds of politico-cultural territorialization and leads them to work out a critical framework for interrogating the practices of "energopolitics". This text itself assumes a "collective assemblage of enunciation", for it brings out the concerns of all blue humanists who are worried about the deteriorating conditions of the sea. It also encourages readers to take into account the enormous significance of the seamless interactions between the sea and the land supported by earth(ing), as far as the decimation of "energopolitical" patterns are concerned.

Chandani Lokugé's *Turtle Nest*

Chandani Lokugé happens to be a renowned Sri Lankan fiction writer who has penned down a number of ground-breaking works including *If the Moon Smiled* (2000), *Turtle Nest* (2003), *Softly, as I Leave You* (2011) and *My Van Gogh* (2020). Born in Sri Lanka, Lokugé works in Australia and is quite aware of the ongoing events in Sri Lanka. Her first novel speaks of the life of Manthri. She is married to Mahendra who is unfortunately not happy with her and accuses her of "treacherous adultery". This novel explores female subjectivity of Manthri who makes attempts to articulate her standpoints while being accused of nuptial disloyalty. Whereas her first work brings out different shades of female subjectivity, *Turtle Nest* teases out the tragic tale of Mala by Aruni who recounts her emotional association with the beaches where her mother, Mala, used to tread. Struggles of Mala to earn a livelihood coupled with the visits of Western tourists to the coastal tourist spots are explored by Aruni. *Softly, as I Leave You* deals with issues pertaining to migration. It examines the sharp tension between togetherness and separateness experienced by the members of a family. *My Van Gogh*, too, takes into account diasporic problems in association with traumatic and post-traumatic complications experienced by Shannon. Within the rich and diverse literary cannon of Lokugé, *Turtle Nest* occupies a singular niche, for it examines some of the contemporary concerns of Blue Humanities.

Scholars have already read *Turtle Nest* from critical points of view. For example, there are scholars who are interested in exploring the poverty-stricken life of Mala, which gets revealed to the readers with the help of Aruni. There are some other scholars who engage themselves in unraveling Aruni's growing sense of belonging to the beach where her mother, Mala, used to dwell in. Diasporic angles in the tale have already been uncovered by some scholars. But, unlike existing critical overtures to Lokugé's fictional narrative, this present study aims to investigate nuanced coastal ecology while interrogating insidious and vicious operations of "energopolitics". In doing so, theoretical recourse to the strands of Blue Humanities is intended to be drawn.

It is unquestionably true that Lokugé's narrative exemplifies complex interactions between energy, ocean and earth(ing) at different turns of the plot. At this point, one may ponder over the implications of the title *Turtle Nest*. Readers are informed at the outset of the fiction that this narrative seeks to revolve around happenings on Lihiniya Island—a territory which is replete with "dozens of baby turtles … [which] hatch under sand, and crawl out" (2017, ix). The depletion of sea turtles followed by the inordinate interventions of Western tourists into the breeding grounds of sea turtles has been seriously addressed in this fiction along with depicting the poignant struggles of Mala. It is quite clear that the free interplay among different energies on the beaches put baby sea turtles at risk and the timely intervention of the fiction writer into the declining conditions of the coastal ecology ultimately turn this into a "minor" text.

This fiction lays bare the fact that tourists from all around the world visit exotic beautiful beaches, intending to experience the rare and endangered marine species which are not available in other parts of the world. Locals dwelling in and around the beaches depend on tourists who would expectedly pay huge money to locals to be taken them to marine species: "The Boys speak in English now. 'We can show you turtles, black and white and brown turtles, turtles laying eggs, and beautiful corals at the bottom of the sea. We take missy and sir in glass-bottom boat?'" (2017, 13). This excerpt suggests that coastal regions stage intersections among capital, human and political energies—the results of which are very ruinous to the sea species including sea turtles. It is true that the sheer poverty of locals is subtly exploited by elite tourists who cause serious damages to endangered sea turtles by getting closer to them. In a way, tourists rely on their capital energy to exploit the straitened lives of locals who are indirectly engaged in the act of taking them to sea turtles. Besides it, the use of "glass-bottomed boat" is a clear reference to how the sea is time and again invaded for recreational purposes. It hints at how crass human activities endanger "corals" which can hardly be reproduced. In a sense, the destruction of sea resources impacts the "intensive" becoming of the sea, and coastal regions that connect terrestrial ecology with marine ecology need to be paid due attention. One may cite some fictional examples to elaborate on how the growth of the tourist industry jeopardizes the geology of beaches in general:

> Once in a way, one or two women might squat nearby to relieve aching of plastic bags full of tourist ware.
>
> (2017, 15)

> The beach boys cluster around. Premasiri has brought a present for Aruni—a beautiful coral ornament shaped like a woman.
>
> (2017, 49)

> ...the beach boys lived off the tourists. They learned their trade from one another, and by fourteen and fifteen, they know all there was to know about drugs, prostitution, pimping and God knows what else.
>
> (2017, 83)

These fictional instances point out how the spread of tourism impacts the depletion of corals and sea turtles. Capital energy is channelized in such a way that "blue trafficking" can take place. More importantly, locals are subtly engaged in committing this blue crime. It also refers to how beaches turn out to be the space for proliferation and dissemination of different blue criminal activities. Tourists attempt to territorialize the coastal regions in terms of energy in such a way that they can get access to "blue archives". The scattering of plastic products on the beach is alarming and detrimental for the good health of sea species as plastics frequently disrupt "geokinesis" of the Earth by means of plasticizing earthly entities. It is true that plastics scattered on the beach do not only ruin the productivity of coastal ecology but also disjoint the "connectivity" and "convergence" between marine ecologies because plastic can only be transformed into different figures. Interestingly, plastic can hardly return to its originality and every time it gets de-constructed, it slips into already always "plastic" state. Every "turn" of plastic is thus actually and necessarily a plastic "re-turn". Therefore, restraints on the use of plastics by tourists could be imposed to destratify "energy flow". Here, one has to understand the function of "energy linkages" that serve to bring sea and land together. Trading activities or drug trafficking require a network of energies which end up affecting the intensive transformation of the sea. Here, one may be reminded of the Sri Lankan Civil War which impacts the inflows of tourists from all over the world:

> As the war went on and on, the rich tourists stopped coming over, so all those others in the fishing village who lived off the tourists went empty handed. See, even these days, missy, how hotels are almost empty, with only the poorer tourists coming in on cheap deals.

(2017, 45)

In a way, political, military and social energies distract the movement of capital energy which has a connection with the protection of coastal ecology from potential human threats. It is obviously true that energies, the sea and coastal ecology are densely interconnected with each other in the sense that dire and dreadful effects of the Civil War deter tourists from visiting. It exacerbates the poverty of locals on the one hand, and it reduces the possibility of the depletion of marine species on the other hand.

This fiction registers how tourists, with the help of locals, go to visit sea turtles in person and put them in danger:

> The turtle is forced upside down. He keeps flapping its short stumps against its inner sides. It's one massive turtle the beach boys hold it down. Premasiri gives them orders and shakes hands and chats with tourists. Aruni asks him to give five hundred rupees to the small boy flitting around like an insect isn't that a lot of money to pay for a glimpse of a fucked-up turtle?

(2017, 120–121)

It clearly divulges how "energopolitics" works in practice, thereby running the risk of endangering sea turtles. The role of Aruni here could be questioned on the grounds that instead of indulging in the configurations of "energopolitics", she could have resisted from doing so the same way she tries hard to establish her maternal "connection" with the land to shed her "tourist" identity. One may take another example to lay bare how "tourists" invade into the spaces of sea turtles—a literary exhibition of how capital energy in association with political, social and cultural energies run into marine energy and subsequently jeopardize sea turtles and other sea animals. In a way, human attempts to territorialize "energy flow" bear dire and dreadful impacts on marine ecologies at large and necessitate critical interventions into the functioning of "energopolitics". It is because of having capital energy that tourists are able to exploit the sea in general and particularly to aid the spree of commercialization across oceanic spaces: "you and Paul sir must come in the glass-bottom boat with us. We'll show you those turtles. They are really something to see. All tourists go for them. We'll give you a special price" (2017, 213–214). This fictional reference clearly substantiates the fact that "energopolitics" is very much at work in oceanic spaces and is undoubtedly a potent threat to the extinction of sea turtles.

Here, one may be likely to argue that as Aruni tries to connect herself with the beach where her mother used to live, she cannot think of getting away from sea turtles, the representatives of the sea. But this contention does not stand to reason because Aruni cannot afford to indulge in the act of being close to sea turtles just because she is in search of her mother's identity. The mysterious silence of Aruni on the involvement of locals as tourist guides needs to be questioned. In fact, when she heard someone saying "A turtle, missy ... It's a turtle laying eggs on the beach, very close. You always wanted to see that. Come, come quickly, it will be gone in a short while" (2017, 221), she "leaned forward now, fully awake, smiling gleefully. She had waited long for this" (2017, 221). This clearly suggests that Aruni consciously indulges in the global tourism industry and stays indifferent to the concerns of Blue Humanities.

Following Aruni's imperviousness to the declining conditions of sea turtles, one may argue that Lokugé has ingeniously portrayed how readers can adopt the concerns of Blue Humanities to develop critical frameworks to question the human efforts to exploit "energy linkages". Here, one may be reminded of the process of "earth(ing)"—a geokinetic movement to elucidate how "energy thinking" can be used as a critical investigative instrument to destratify the "mattering" of energy in general and to delimit the jurisdiction of Blue Humanities to the extent of coastal ecology. At one point in the narrative, Lokugé brings out Aruni's deep-seated desire to be a "mother turtle" so that she can both keep "contact with everyone on the beach" and dwell "at the bottom of the sea" (2017, 18). This fictional excerpt places emphasis on the huge importance of "coastal ecology" which needs to be enfolded into Blue Humanities while formulating "energy thinking" as a tool of critical enquiry. Here, Aruni wishes to be united with the seamless interactions between land and sea on the beach. She, as it were, renders the Earth a body without organs (BwO) that stands interspersed with diagrammatic and rhizomatic trajectories of energy.

Energy Thinking: A Way Forward

It has now become quite clear that select "minor" narratives represent the "assemblages" of energy, ocean and earth(ing) in different ways. But, interestingly, it may be noted that it is by showing the complex and nuanced intersections among energy, ocean and earth(ing), authors of select "minor" writings call for considering the epistemic strands of remapped Blue Humanities to combat the strikes of "energopolitics" at praxis. In this regard, "energy thinking"[2] can tenably be worked out in such a way that, just as "energy linkages" seek to explain how the "matter" and "mattering" of energy condition the geological becoming of the Earth, "energy thinking" could be resorted to elucidate how the "geokinetic" movement of the Earth in general is constantly fuelled by the supply and transduction of energies; how energies travel through thoughts while constructing and deconstructing the geology of matter; how energies form, de-form and transform the world around us through human and non-human actions and interactions; how human efforts to regulate "energy flows" result in ecological catastrophes; and so on and so forth. "Energy thinking" actually implies the pervasive and inclusive operational logic by means of which geological ungrounding of the Earth, materiality of energy and intensive differentiality of ocean can be figured out. Apart from this, "energy thinking" also helps one decipher how energy seamlessly links and delinks, connects and disconnects and joins and disjoins territory and the Earth.

It is also true that select "minor" writings pave the way for the functioning of "energy thinking" at different levels and to different degrees. For example, *Reef* underscores how "energy thinking" could be resorted to to take the measure of the appalling ramifications of having "energy infrastructure" in terms of a "marine park" (1994, 177) and "floating restaurant" (1994, 177) in the sea. *Swimming in the Monsoon Sea* demonstrates how "energy thinking" could be used to delve deeper into the imminent upshots of "plastiglomerate" which comprises "a mixture of plastic-intermateriality – surprising, bizarre, becoming, erratic, and aberrant" (Ghosh, "The Plastic Controversy", 2021). *Mosquito* too points at how "energy thinking" can be of some help to gauge the dire and dreadful consequences of having "a huge high-rise hotel" near the beach and to take into account the fact that like "memory, the sea had a life of its own" (2007, 265). *Turtle Nest*, lays bare how "energy thinking" can lead one to take stock of how the eggs laid by sea turtles in the full-moon nights stand "exposed to thieves" (2017, 222). It suggests that human capitalistic intention to turn the eggs of sea turtles into capital energies triggers disappearance and depletion of sea turtles from their natural habitats: "Even the turtle seems to have disappeared…" (2017, 223). Thus, it is clear that the operative logic of "energy thinking" is contingent upon dual axes of "energy linkage" and "energy diagram", and the select Sri Lankan "minor" writings attest to this contention.

Notes

1 In "What is an Assemblage?", Thomas Nail explains how an assemblage works by the principle of "deterritorialization" and facilitates one to put forward "revolutionary aims":

> all assemblages are constantly changing according to four different kinds of change or "deterritorialization … all assemblages are political. If we want to know what an assemblage is, we need to know how it works. Once we understand how the assemblage functions, we will be in a better position to perform

diagnosis: to direct or shape the assemblage toward increasingly revolutionary aims.

(2017, 37)

2 In *Trans(in)fusion: Reflections for Critical Thinking*, Ghosh pertinently contends:

Thinking theory is thinking energy. Is energy really a thought? Or does it stay as "thinking" in the process of uncovering a thought? Energy is, for me, a challenge to "identity thinking"; it is revealed as the "non-identical", which is why, energy comes home to get interpreted as the aesthetics of "counter". (2021, 43). This critical reflection calls for "energy thinking" as a mode of critical investigation to actualize "the aesthetics of 'counter'". In short, "energy thinking" seems to facilitate one to pursue the aleatory "points of departure" in the process of "uncovering a thought".

Conclusion

How Energy Humanities Matter

Energy is generally understood to be a matter of interdisciplinary science studies and thus numerous attempts are being made to map how the flows of energy can be channelized and re-channelized to impact the flows of the economy through global markets. Opposed to this traditional understanding of energy in terms of science studies, there is a need to turn to the humanities so far as social, political, cultural, economical and other associated dimensions of energy are concerned. In other words, mathematical reasoning can help an individual figure out different measurements of energy flows whereas humanitarian perspectives, in particular, seek to offer how energy does not stand confined in the boundaries of science studies and more importantly it affects individuals' ways of being in the postglobalized societies. Whereas Foucault laid down the concept of biopolitics to take stock of how the government resorts to a number of administrative measures to put a check on the political, social and cultural activities of individuals, Gokce Gunel in "Ergos: A New Energy Currency" holds energopolitics to be an extension of "disciplinary biopolitics" (2014, 365), thereby implying that energopolitics is a state-monitored biopolitics, that helps a government keep an eye on the movements of human and nonhuman entities on Earth. In short, energopolitics ends up being held as a monitoring tool to exercise governmental surveillance on human and nonhuman bodies. Keeping that in mind, it can be argued that the epistemic limits of energopolitics call for the refashioning of human engagement with energy flows on ethical grounds so that political exploitations of energy flows supported by governmental authority can be contained to a great extent. This contention can further be worked out this way that human beings need to be trained in forging ethically sound engagements with the uses of energy so that people across borders and boundaries can have equal access to different energy resources.

At this point, one may stop and ponder the following questions: Who has the power to determine one's access to diverse energy resources? Does one's access to energy resources depend on his economic conditions and political affiliation? How can the apparent dichotomy between energy production and energy supply be resolved? Whereas in "Beyond Oil: The Emergence of the

DOI: 10.4324/9781032629728-7

Energy Humanities", Jamie L. Jones argues: "Energy humanities is a new field" (2019, 156), Leo Coleman spells out in "Afterword: People Thinking Energetically" how energy matters to our personal and political lives: "Energy is, at once, personal, collective and political, an experienced reality" (2021, 181). Coleman seems to suggest that there is a need to turn to Energy Humanities that speak of how individuals need to be trained in thinking with the operative logic of energy so that energy can be employed for the positive changes of the society and more importantly, an ethical re-engagement with energy needs to be mapped to set "energy thinking" in motion.

It is undoubtedly true that energopolitics entails the spread of energy discrimination at the local level, that could be countered by means of reorganizing "energy thinking" in ethical terms. It means that energy supply and energy consumption need to be mapped in government policies so that access to energy resources could be extended to one and all dwelling on the Earth.

Usually, developed nations, by virtue of possessing economic power, tend to engage in energopolitical activities to customize one's access to energy resources, taking local geopolitical factors into account. Opposed to this energopolitical narrative, thinking with the operative logic of energy backed by the perspectives from the marginalized people and geokinetic becomings of the Earth needs to be epistemized to call energy discrimination into question and to facilitate one's access to multiple energy resources. Government needs to adoptp a number of initiatives to actualize the visions of Energy Humanities, that is, to liberate energy flows from the pull of any territorialized epistemological overture and to let everyone recognize *figural-functional* dynamics of energy.

In the context of postcolonial studies, does "energy humanities" have anything to contribute? In other words, can critical interventions pertaining to "energy humanities" be taken into account to upgrade and update postcolonial inroads into ecological disasters happening in contemporary times? These questions can be responded to by means of arguing that the transfer of energy plays a crucial role in channelizing power politics in the zones of hybridity and, more importantly, "energy thinking" from the perspectives of the marginalized, could be weaponized against the energopolitics-led neocolonial practices that seek to have economic authority on fiscal activities of so-called Third World nations. *Figural-functional* dynamics of energy could be employed to put up postcolonial resistance to manipulative uses of energy for having political authority over the margins. Energy Humanities has to work towards liberating the flow of energy from any form of territorialization and to contribute to prevent postcolonial insights from being subjected to a series of stratified categorizations. Nada Kujundžíc and Matúš Mišík have rightly reflected in "Conclusion: Where We Are and Where We Are Going": "The energy humanities is especially well positioned to study both the minutiae of the energy transition, a complex process that impacts all aspects of human life, and the negative effects of a 'business-as-usual' approach to GHG emissions and climate change" (2021, 201). They are of this standpoint that poetics and politics of nomadic singularities of energy can play an important role in empowering postcolonial insights to better deal with the issues related to climate change and planetary crisis. The world at present stands at risks as climatic catastrophes increase in number and exacerbate supposed ecological stability between human and nonhuman beings. Epistemic grafting of "energy thinking" onto the existing and emergent postcolonial insights prove to be effective in the sense that it can both enlarge the scope of "energy humanities" in the domain of planetarity and can enable postcolonial scholars to think about how the flow of energy could be prioritized to question neoliberal exploitations of "energy resources" found across the Earth.

Being an epistemic offshoot of biopolitics, "energopolitics", as it has been conceived by Boyer, gets depicted as being a form of state politics—against which energy ethics could be posited to empower postcolonial resistance to neoliberal economic overtures led by developed nations. At this point, one may stop and think: What is energy ethics? How can energy ethics be used to delimit energy flow in the context of marauding marches of neoliberalism? Energy ethics could be understood as a referent to the ethical use of energy in practice. This means that postcolonial thinkers need to ponder over the possible ways through which energy could ethically be negotiated to construct discourses of counter-strikes. Put in other words, an ethical engagement with energy is required to make calculative and systematic use of energy resources in praxis. In order to actualize energy ethics, invasive and pervasive strands of neoliberalism could be brought under scrutiny and subsequently, energy thinking needs to be tied up with energy ethics to pinpoint the loopholes of neoliberalism in maintaining ecological balance between human and nonhuman beings in reality. This contention can further be worked out in this way so that micropolitics and micropoetics of energy need to be considered together with energy thinking and energy ethics to make significant advancements in the domain of Energy Humanities. One cannot but agree with Helmar Krupp who aptly holds in *Energy Politics and Schumpeter Dynamics*:

> Energy is the physical fuel of societal dynamics. Its form and use is an integral part of a society. Therefore, the politics and the economics of energy provision and consumption are too complex to be left to any particular discipline or elite. Also, there is no such discipline as energology.
> (1992, 11).

This clearly reflects that Energy Humanities has quite good reasons to matter as long as earth(ing) exists.

Energy Humanities could well be extended to epistemic configurations of different variants of posthumanism not only because energy works as a mediator between earthly bodies but the flow of energy could be regulated in different ways to spell out human engagement with nonhuman entities surrounded by technological advancements. For example, advancements in the field of technology could be subjected to energy thinking in the sense that it can help one figure out how the manipulation of energy with the assistance of technology can help in governing the movements of the human and nonhuman agents of the Earth. In other words, transhumanism could be intervened in terms of technological control of energy flow that plays massive role in redefining human-technology interface. In addition, within the domain of Spirituality Studies, energy thinking can play a devisive role in helping one figuring out how energy connects a human being with God in the transcendental world. One may be reminded of "Humans Have Always Been Posthuman: A Spiritual Genealogy of Posthumanism" where Francesca Ferrando reflects: "Pluralistic monism, or monistic pluralism, can be accessed through physics, when considering that many dimensions may exist, each depending on different vibrations of quantum loops of energy called strings" (2016, 247). She is right in holding the opinion that energy bridges the "human" with the "posthuman" condition of being thereby leading one to the experiences of spiritual salvation. In fact, one may also find parallel references in Indic epistemological traditions where energy thinking has taken a central position and seeks to impact Indic paths to spiritual awakening. For example, Tantric *sādhakas* attempt to find different spiritual-material means to channelize energy flows in particular fashions so as get connected with the transcendental Other. In "Reconfiguring Asian Modernity: Negotiating Tantric Epistemological Traditions", Abhisek Ghosal and Bhaskarjyoti Ghosal have extensively contended that the spiritual-material modes of exploration of the interface between inner being and outer being are premised on how a Tantric *sādhaka* engages with transduction of energy through different bodies: "Tantric sādhakas intend to tap into 'energy linkages' to switch from one body to another. They take part in the complex process of worlding and have strong convictions in the evolving nature of Tantra" (2022, 37).

Finally, to conclude the discussion on the relevance of Energy Humanities in contemporary times, it can be argued that an understanding of the relevance of energy thinking in figuring out contemporary issues and challenges pertaining to socio-cultural, political, ecological and economical alterations has to be in tune with how an individual engages with energy perspectives which are always being fine-tuned by the scholars working in the field of Energy Studies. It means that epistemic contours of Energy Humanities are quite porous; therefore, it invites scholars working in different interdisciplinary studies to come and contribute to the advancements of energy thinking. The importance and popularity of Energy Humanities are going to beaugmented in Science Studies not just because energy cuts across humanities and sciences transversally but also because energy is a condition for the actualization of intensive differentialities that constitute the onto-epistemological transformations of existing disciplines and discourses.

References

Alaimo, Stacy. 2010. *Bodily Natures: Science, Environment and the Material Self.* Bloomington: Indiana University Press.

Alaimo, Stacy. 2019. "Introduction: Science Studies and the Blue Humanities." *Configurations*, 27 (4): 429–432. https://doi.org/10.1353/con.2019.0028.

Anderson, Jon and Kimberley Peters. 2014. "A Perfect and Absolute Blank': Human Geographies of Water Worlds." In *Water Worlds: Human Geographies of the Ocean*, edited by Jon Anderson and Kimberley Peters, 3–19. Burlington: Ashgate.

Arnold, V.I. 1992. *Catastrophe Theory.* Berlin: Springer.

Ayres, Robert. 2016. *Energy, Complexity and Wealth Maximization.* Switzerland: Springer.

Bartels, Anke Lars Eckstein, Nicole Waller, and Dirk Wiemann. 2019. *Postcolonial Literatures in English: An Introduction.* Germany: J. B. Metzler Verlag.

Barthes, Roland. 1957. *Mythologies.* New York: The Noonday Press.

Barthélémy, Jean-Hugues. 2012. "Fifty Key Terms in the Works of Gilbert Simondon." In *Gilbert Simondon: Being and Technology*, edited by Arne De Boever, Alex Murray, Jon Roffe, and Ashley Woodward, 203–231. Edinburgh: Edinburgh University Press.

Barrad, Karen. 2007. *Meeting the Universe Halfway: Quantum Physics and the Entanglement of Matter and Meaning.* Durham: Duke University Press.

Bear, Christopher. 2019. "The Ocean Exceeded: Fish, Flows and Forces." *Dialogues in Human Geography* 9 (3): 329–332. DOI: 10.1177/2043820619878567.

Bensmaia, Reda. 1986. "Foreword: The Kafka Effect by Réda Bensmaïa." In *Kafka: Toward a Minor Literature*, edited by Gilles Deleuze and Felix Guattari, ix–xxi. Minneapolis: University of Minnesota Press.

Bennett, Jane. 2010. *Vibrant Matter: A Political Ecology of Things.* Durham: Duke University Press.

Bhattacharyya, Susmita. 2015. "Parallels in Physics and Philosophy." *Prācyā* 7 (1): 38–43. https://www.pracyajournal.com/article/48/7-1-4-349.pdf.

Bloomfield, Mandy. 2019. "Widening Gyre: A Poetics of Ocean Plastics." *Configurations* 27 (4): 501–523. https://doi.org/10.1353/con.2019.0033.

Blum, Hester. 2010. "The Prospect of Oceanic Studies." *PMLA* 125 (3): 670–677. https://www.jstor.org/stable/25704464.

Bogue, Ronald. 1997. "Minor Writing and Minor Literature." *Symploke* 5(1/2): 99-118. https://www.jstor.org/stable/40550404.

Boyer, Dominic. 2014. "Energopower: An Introduction." *Anthropological Quarterly* 87 (2): 309–334.

Boyer, Dominic. 2019. *Energopolitics: Wind and Power in the Anthropocene*. Durham: Duke University Press.

Brand, Dionne. 2001. *A Map to the Door of No Return: Notes on Belonging*. Toronto: Doubleday Canada.

Brewer, Jennifer. 2017. "Actualizing Marine Policy Engagement." *Dialogues in Human Geography* 7 (1): 45–49. DOI: 10.1177/2043820617691648.

Buchanan, Ian. 2021. *Assemblage Theory and Method*. London: Bloomsbury Academic.

Buchanan, Ian. 2019. "Must We Eat Fish?" *Symploke* 27 (1–2): 79–90. https://muse.jhu.edu/article/734652.

Buchanan, Ian and Celina Jeffery. 2019. "Toward a Blue Humanity." *Symploke* 27 (1–2): 11–14. https://muse.jhu.edu/article/734647.

Carson, Rachel. 1961. *The Sea Around Us*. New York: Oxford University Press.

Caruth, Cathy. 1996. *Unclaimed Experience: Trauma, Narrative and History*. Baltimore: John Hopkins University Press.

Castoriadis, Cornelius. 1997. *The Castoriadis Reader*. Translated by David Ames Curtis. Oxford: Blackwell.

Chakrabarty, Dipesh. 2021. *The Climate of History: In a Planetary Age*. Chicago: The University of Chicago Press.

Chattopadhyay, Shrimanta. n.d. "Rgvedic *Dyāvāpṛthivī* and Modern Science: A Relevant Study." Unpublished Manuscript. Typescript.

Choi, Young Rae. 2017. "The Blue Economy as Governmentality and the Making of New Spatial Rationalities." *Dialogues in Human Geography* 7 (1): 37–41. DOI: 10.1177/2043820617691649.

Colebrooke, Claire. 2022. "Geophilosophy as the End of Philosophy." *Subjectivity* 15: 169–186. https://doi.org/10.1057/s41286-022-00136-5.

Coleman, Leo. 2021. "Afterword: People Thinking Energetically." In *Ethnographies of Power: A Political Anthropology of Energy*, edited by Tristan Loloum, Simone Abram, and Nathalie Ortar, 180–194. New York: Berghahn.

Conca, Ken. 2016. "The Changing Shape of Global Environmental Politics." In *New Earth Politics: Essays from the Anthropocene*, edited by Simon Nicholson and Sikina Jinnah, 21–42. Massachusetts: The MIT Press.

Crist, Eileen. 2019. *Abundant Earth: Toward an Ecological Civilization*. Chicago: The University of Chicago Press.

Daimari, Esther. 2017. "Landscape in Romesh Gunesekera's *Reef, The Sandglass*, and *Heaven's Edge*." *South Asian Review* 38 (2): 49–64. DOI:10.1080/02759527.2017.12002560.

Deleuze, Gilles. 1994. *Difference and Repetition*. Translated by Paul Patton. New York: Columbia University Press.

Deleuze, Gilles. 1993. *The Fold: Leibniz and the Baroque*. Translated by Tom Conley. London: The Athlone Press.

Deleuze, Gilles and Felix Guattari. 1987. *A Thousand Plateaus: Capitalism and Schizophrenia*. Translated by Brian Massumi. Minneapolis: University of Minnesota Press.

Deleuze, Gilles and Felix Guattari. 1983. *Anti Oedipus: Capitalism and Schizophrenia*. Translated by Robert Hurley, Mark Seem, and Helen R. Lane. Minneapolis: University of Minnesota Press.

Deleuze, Gilles and Felix Guattari. 1986. *Kafka: Toward a Minor Literature*. Translated by Dana Polan. Minneapolis: University of Minnesota Press.

Deleuze, Gilles and Felix Guattari. 1994. *What is Philosophy? Translated by Hugh Tomlinson and Graham Burchell.* New York: Columbia University Press.

DeLoughrey, Elizabeth M. 2019a. *Allegories of the Anthropocene.* Durham: Duke University Press.

DeLoughrey, Elizabeth M. 2007. *Routes and Roots: Navigating Caribbean and Pacific Island Literatures.* Honolulu: University of Hawai Press.

DeLoughrey, Elizabeth M. 2019b. "Toward a Critical Ocean Studies for the Anthropocene." *English Language Notes* 57 (1): 21–36. https://muse.jhu.edu/article /724603.

Derrida, Jacques. 2008. *The Animal That Therefore I Am.* Edited by Marie-Luise Mallet. New York: Fordham University Press.

Derrida, Jacques. 1978. *Writing and Difference.* Translated by Alan Bass. Chicago: The University of Chicago Press.

Derrida, Jacques and Eric Prenowitz. 1995. "Archive Fever: A Freudian Impression." *Diacritics* 25 (2): 9–63. http://www.jstor.org/stable/465144.

Discover Sedge South Africa. 2022. *Ocean Trauma.* https://www.discover-sedgefield -south-africa.com/.

Dobrin, Sidney I. 2021. *Blue Ecocriticism and the Oceanic Imperative.* New York: Routledge.

Dora, Veronica Della. 2021. *The Mantle of the Earth: Genealogies of a Geographical Metaphor.* Chicago: The University of Chicago Press.

Duffy, Simon B. 2013. *Deleuze and the History of Mathematics: In Defense of the "New".* London: Bloomsbury.

Dukes, Hunter. 2016. "Assembling the Mechanosphere: Monod, Althusser, Deleuze and Guattari." *Deleuze and Guattari Studies* 10 (4): 514–530. DOI: 10.3366/ dls.2016.0243.

Earle, Sylvia A. 2010. *The World Is Blue: How Our Fate and the Ocean's Are One.* Washington: National Geographic Society.

Ferrando, Francesca. 2016. "Humans Have Always Been Posthuman: A Spiritual Genealogy of Posthumanism." In *Critical Posthumanism and Planetary Futures*, edited by Debashish Banerji and Makarand R. Paranjape, 243–256. London: Springer.

Foucault, Michel. 2003. *"Society Must Be Defended."* Translated by David Macey. New York: Picador.

Freeden, Willi. 2015. "Geomathematics: Its Role, Its Aim, and Its Potential." In *Handbook of Geomathematics*, edited by Willi Freeden, M. Zuhair Nashed, and Thomas Sonar, 4–79. London: Springer.

Gale, Ken. 2016. "Writing Minor Literature: Working With Flows, Intensities and the Welcome of the Unknown." *Qualitative Inquiry* 22 (5): 301–308. DOI: 10.1177/1077800415615615.

Geerts, Robert-Jan, Bart Gremmen, Josette Jacobs, and Guido Ruivenkamp. 2014. "Towards a Philosophy of Energy." *ScientiæStudia* 5 (12): 105–127. http://dx.doi .org/10.1590/S1678- 31662014000400006.

Ghosal, Abhisek. 2023. *Plasti(e)cological Thinking: Working out an (Infra)structural Geoerotics.* Spain: Vernon Press.

Ghosal, Abhisek and Bhaskarjyoti Ghosal. 2022. "Reconfiguring Asian Modernity: Negotiating Tantric Epistemological Traditions." *Symploke* 30 (1–2): 33–46. https://muse-jhu-edu-christuniversity.knimbus.com/article/885929.

Ghosh, Ranjan. 2012. "Globing the Earth: The New-logics of Nature." *SubStance* 41 (1): 3–14. https://doi.org/10.1353/sub.2012.0008.

Ghosh, Ranjan. 2019. "Plastic Literature." *University of Toronto Quarterly* 88 (2): 277–291. https://muse.jhu.edu/article/732562.

Ghosh, Ranjan. 2021. "The Plastic Controversy." *Critical Inquiry*, Feb 4. https://critinq .wordpress.com/2021/02/04/the-plastic-controversy/.

Ghosh, Ranjan. 2021. *Trans(in)fusion: Reflections for Critical Thinking*. London: Routledge.

Griffith, Ralph T. H., trans. 1896. *The Hymns of the Rigveda*. Benares: E.J. Lazarus.

Guattari, Felix. 2000. *The Three Ecologies*. London: The Athlone Press.

Gunel, Gokce. 2014. "Ergos: A New Energy Currency." *Anthropological Quarterly* 87 (2): 359–379. http://dx.doi.org/10.1353/anq.2014.0026.

Gunesekera, Romesh. 1994. *Reef*. London: Granta Books.

Han, Lisa. 2019. "The Blue Frontier: Temporalities of Salvage and Extraction at the Seabed." *Configurations* 27 (4): 463–481. https://doi.org/10.1353/con.2019.0031.

Hernes, Tor. 2014. *A Process Theory of Organization*. Oxford: Oxford University Press.

Herzogenrath, Bernd. 2013. "White." In *Prismatic Ecology: Ecotheory beyond Green*, edited by Jeffrey Jerome Cohen, 1–21. Minneapolis: University of Minnesota Press.

Honarpisheh, Donna. 2019. "The Sea as Archive: Impressions of *Qui Se Souvient De La Mer*." *Symploke* 27 (1–2): 91–109. https://muse.jhu.edu/article/734653.

Huber, Matthew T. 2018. "Fossilized Liberation: Energy, Freedom, and the 'Development of the Productive Forces.'" In Chapter 17 of *Materialism and the Critique of Energy*, edited by Brent Ryan Bellamy and Jeff Diamanti, 501–524. Chicago: M.C.M.

Jacobs, Stanley. 1989. "A Philosophy of Energy." *Holistic Medicine* 4 (2): 95–111. DOI: 10.3109/13561828909023024.

Jazeel, Tariq. 2003. "Unpicking Sri Lankan 'island-ness' in Romesh Gunesekera's *Reef*." *Journal of Historical Geography* 29 (4): 582–298. doi:10.1006/jhge.2002.0410.

Johnson, Marcha. 2016. "Introduction." In *Coastal Change, Ocean Conservation and Resilient Communities*, edited by Marcha Johnson and Amanda Bayley, 1–6. Switzerland: Springer.

Jones, Andrew. 2007. *Memory and Material Culture*. Cambridge: Cambridge University Press.

Jones, Christopher F. 2018. "The Materiality of Energy." *Canadian Journal of History* 53 (3): 378–394. https://muse.jhu.edu/article/714234.

Jones, Jamie L. 2019. "Beyond Oil: The Emergence of the Energy Humanities." *Resilience: A Journal of Environmental Humanities* 6 (2–3): 155–163. https://doi .org/10.5250/resilience.6.2-3.0155.

Jue, Melody. 2020. *Wild Blue Media: Thinking through Seawater*. Durham: Durham University Press.

Kant, Amitabh, Pramit Dash, and Piyush Prakash. 2022. "Leapfrogging the Indian Blue Economy." *Employment News*, March 12.

Kaushik, Madhuchanda. 2015. "Vedic Thought and Modern Science." *Prācyā* 7 (1): 138–144. https://www.pracyajournal.com/article/61/7-1-18-801.pdf.

Kinsley, David R. 1988. *Hindu Goddesses: Visions of the Divine Feminine in the Hindu Religious Tradition*. California: University of California Press.

Kleinstreuer, Clement. 2010. *Basic Theory and Selected Applications in Macro- and Micro- Fluidics*. New York: Springer.

Krupp, Helmar. 1992. *Energy Politics and Schumpeter Dynamics*. Japan: Springer.

Kujundži´c, Nada and Matúš Mišík. 2021. "Conclusion: Where We Are and Where We Are Going." In *Energy Humanities. Current State and Future Directions*, edited by Matúš Mišík and Nada Kujundžić, 199–204. Switzerland: Springer.

Kumar, Abhinandan. 2016. "Energy Generation Through Vedas." *International Journal of Science Technology and Management* 5 (3): 294–304.

Latour, Bruno. 2005. *Reassembling the Social: An Introduction to Action Network Theory*. Oxford: Oxford University Press.

Lewis, Martin W. and Karen E. Wigen. 1997. *The Myth of Continents: A Critique of Metageography*. California: University of California Press.

Lokugé, Chandani. 2017. *Turtle's Nest*. North Melbourne: Arcadia.

Lovelock, James. 2004. "Reflections on Gaia." In *Scientists Debate Gaia: The Next Century*, edited by Stephen H. Schneider, James R. Miller, Eileen Crist, and Penelope J. Boston, 1–5. Massachusetts: The MIT Press.

Luc-Nancy, Jean. 2000. *Being Singular Plural*. California: Stanford University Press.

Malabou, Catherine. 2009. *Ontology of the Accident: An Essay on Destructive Plasticity*. Translated by Carolyn Shread. Cambridge: Polity.

Malabou, Catherine. 2005. *The Future of Hegel: Plasticity, Temporality and Dialectic*. New York: Routledge.

Marche, Guillaume. 2012. "Introduction Why Infrapolitics Matters." *Revue francaise d'etudes americaines* 131 (1): 3–14. http://dx.doi.org/10.3917/rfea.131.0003.

Marder, Michael. 2017. *Energy Dreams: Of Actuality*. New York: Columbia University Press.

Mbembe, Achille. 2002. "The Power of the Archive and its Limits." In *Refiguring the Archive*, edited by Carolyn Hamilton, Veme Harris, Jane Taylor, Michele Pickover, Graeme Reici, and Razia Saleh, 19–26. Dordrecht: Springer.

Mentz, Steve. 2024. *An Introduction to the Blue Humanities*. London: Routledge.

Mentz, Steve. 2019b. "Shakespeare and Blue Humanities." *SEL Studies for English Literature 1500–1900* 59 (2): 383–392. https://doi.org/10.1353/sel.2019.0018

Mickey, Sam. 2016. *Whole Earth Thinking and Planetary Coexistence Ecological Wisdom at the Intersection of Religion, Ecology, and Philosophy*. London: Routledge.

Mišík, Matúš and Nada Kujundžić, eds. 2021. *Energy Humanities. Current State and Future Directions*. Switzerland: Springer.

Moore, J., ed. 2016. *Anthropocene or Capitalocene? Nature, History, and the Crisis of Capitalism*. Binghamton: PM Press.

Mukherjee, Jenia. 2020. *Blue Infrastructures: Natural History, Political Ecology and Urban Development in Kolkata*. Singapore: Springer.

Mukhopadhyay, Anway. 2018. *The Goddess in Hindu-Tantric Traditions: Devi as Corpse*. London: Routledge.

Mukhopadhyay, Ramaranjan. 2001. *Rasa Samīkṣā*. Kolkata: Sanskrit Pustaka Bhander.

Nail, Thomas. 2021. *Theory of the Earth*. California: Stanford University Press.

Nail, Thomas. 2017. "What is an Assemblage?" *SubStance* 46 (1): 21–37. https://muse.jhu.edu/article/650026.

Nancy, Jean-Luc. 2000. *Being Singular Plural*. California: Stanford University Press.

Neimanis, Astrida. 2017. *Bodies of Water: Posthuman Feminist Phenomenology*. London: Bloomsbury.

Neyrat, Frederic. 2019. *The Unconstructable Earth: An Ecology of Separation*. New York: Fordham University Press.

Nicholson, Simon and Sikina Jinnah. 2016. "Living on a New Earth." In *New Earth Politics: Essays from the Anthropocene*, edited by Simon Nicholson and Sikina Jinnah, 1–16. Massachusetts: The MIT Press.

Nordenson, Catherine Seavitt, Guy Nordenson, and Julia Chapman. 2018. *Structures of Coastal Resilience*. New York: Island Press.

Oppermann, Serpil. 2019. "Storied Seas and living Metaphors in the Blue Humanities." *Configurations* 27 (4): 443–461. https://doi.org/10.1353/con.2019.0030.

Parr, Adrian. 2018. *Birth of a New Earth: The Radical Politics of Environmentalism*. New York: Columbia University Press.

Pearson, Michael. 2003. *The Indian Ocean*. London: Routledge.

Perez, Craig Santos. 2020. "The Ocean in Us: Navigating the Blue Humanities and Diasporic Chamoru Poetry." *Humanities* 9 (3): 1–11. doi:10.3390/h9030066.

Peters, J.D. 2015. *The Marvelous Clouds: Toward a Philosophy of Elemental Media*. Chicago: University of Chicago Press.

Probyn, Elspeth. 2016. *Eating the Ocean*. Durham: Durham University Press.

Quigley, K. 2023. *Reading Underwater Wreckage: An Ecrusting Ocean*. London: Bloomsbury.

Rapp, Bastian E. 2017. *Microfluidics: Modeling, Mechanics and Mathematics*. London: Elsevier Inc.

Rawes, Peg. 2008. *Space, Geometry and Aesthetics: Through Kant and Towards Deleuze*. London: Palgrave Macmillan.

Ruiz, Rafico and Melody Jue eds. 2021. *Saturation: An Elemental Politics*. Durham: Durham University Press.

Sagan, Dorion and Lynn Margulis. 1997. "Gaia and Philosophy." In *Slanted Truths: Essays on Gaia, Symbiosis and Evolution*, edited by Lynn Margulis and Dorion Sagan, 145–147. New York: Springer-Verlag.

Sauvagnargues, Anne. 2016. *ARTMACHINES: Deleuze, Guattari, Simondon*. Translated by Suzanne Verderber and Eugene W. Holland. Edinburgh: Edinburgh University Press.

Scott, Harry Pitt. 2020. "Offshore Mysteries, Narrative Infrastructure: Oil, Noir, and the World-Ocean." *Humanities* 9 (71): 1–15. doi:10.3390/h9030071.

Sentesy, Mark. 2020. *Aristotle's Ontology of Change*. Evaston: Northwestern University Press.

Selvadurai, Shyam. 2005. *Swimming in the Monsoon Sea*. Toronto: Tundra Books.

Serres, Michel. 2017. *Geometry: The Third Book of Foundations*. Translated by Randolph Burks. London: Bloomsbury.

Serres, Michel. 1995. *The Natural Contract*. Translated by Elizabeth MacArthur and William Paulson. New York: University of Michigan Press.

Sharma, R.S. 2020. "Amitav Ghosh, Indian Ocean and the New Thalassology." In *Indian Ocean: The New Frontier*, edited by Kousar Azam, 131–139. London: Routledge.

Silva, Neluka. 2012. "'No Place Called Home?': Representations of Home in Chandani Lokugé's *If the Moon Smiled* and Roshi Fernando's *Homesick*." *South Asian Review* 33 (3): 109–123. DOI: 10.1080/02759527.2012.11932898.

Smil, Vaclav. 2003. *Energy at the Crossroads: Global Perspectives and Uncertainties.* Cambridge: The MIT Press.

Smith, Daniel W. 2008. "Deleuze and the Production of the New." In *Deleuze, Guattari and the Production of the New*, edited by Simon O'Sullivan and Stephen Zepke, 151–161. New York: Continuum.

Smith, Jones. 2018. "Medieval Water Energies: Philosophical, Hydro-Social, and Intellectual." *Open Library of the Humanities* 4 (2):1-28 doi: https://doi.org/10.16995/olh.228.

Smith, James L. and Steve Mentz. 2020. "Learning an Inclusive Blue Humanities: Oceania and Academia through the Lens of Cinema." *Humanities* 9 (67): 1–14. doi:10.3390/h9030067.

Steinberg, Philip E. 2013. "Of Other Seas: Metaphors and Materialities in Maritime Regions." *Atlantic Studies* 10 (2): 156–169. http://dx.doi.org/10.1080/14788810.2013.785192.

Steinberg, Philip E. 2001. *The Social Construction of the Ocean.* Cambridge: Cambridge University Press.

Steinberg, Philip E. and Kimberley Peters. 2015. "Wet Ontologies, Fluid Spaces: Giving Depth to Volume through Oceanic Thinking." *Environment and Planning D: Society and Space* 33: 247–264. https://journals.sagepub.com/doi/pdf/10.1068/d14148p?casa_token=e3TcKE_5pNsAAAAA:zf6Q_TOksw5cn535w5Vk9AhRHubzDRCCtAs3ZGdGYM_PS7AcmcRkUaM6WkhV_ddZ7Cl3urghBFtnmA.

Stoekl, Allan. 2018. "Marxism, Materialism, and the Critique of Energy." In Chapter 1 of *Materialism and the Critique of Energy*, edited by Brent Ryan Bellamy and Jeff Diamanti, 1–28. Chicago: M.C.M.

Stoner, Jill. 2012. *Toward A Minor Architecture.* Cambridge: The MIT Press.

Subramanian, Lakshmi. 2020. "India and the Indian Ocean: Old Concerns and New Perspectives." In *Indian Ocean: The New Frontier*, edited by Kousar Azam, 65–82. London: Routledge.

Szeman, Imre. 2014. "Conclusion: On Energopolitics." *Anthropological Quarterly* 87 (2): 453–464. https://muse.jhu.edu/article/545595.

Szeman, Imre. 2021. "Towards a Critical Theory of Energy." In Chapter 2 of *Energy Humanities*. Current State and Future Directions, edited by Matúš Mišík and Nada Kujundžić, 23–36. Switzerland: Springer.

Szerszynski, Bronislaw and John Urry. 2010. "Changing Climates: Introduction." *Theory, Culture & Society* 27 (2–3): 1–8. DOI: 10.1177/0263276409362091.

Tally Jr., Robert T. 2011. "Preface: The Timely Emergence of Geocriticism." In *Geocriticism: Real and Fictional Spaces*, edited by Bertrand Westphal, ix–xiii. New York: Palgrave Macmillan.

Tally Jr., Robert T. and Christine Battista. 2016. "Introduction: Ecocritical Geographies, Geocritical Ecologies, and the Spaces of Modernity." In *Ecocriticism and Geocriticism: Overlapping Territories in Environmental and Spatial Literary Studies*, edited by Robert T. Tally and Christine Battista, 1–15. New York: Palgrave Macmillan.

Tearne, Roma. 2007. *Mosquito.* London: HaperCollins Publishers.

Te Punga Somerville, A. 2017. "Where Oceans Come From," *Comparative Literature* 69 (1): 25–31.

Tyrrell, Toby. 2013. *A Critical Investigation of the Relationship between Life and Earth*. Princeton: Princeton University Press.

Vartabedian, Becky. 2018. *Multiplicity and Ontology in Deleuze and Badiou*. Switzerland: Palgrave Macmillan.

Vaughan, Mason and Alden T. Vaughan, eds. 2011. *The Tempest*. London: Bloomsbury.

Wang, Fred. 2021. "Ocean Memory: Humans have Memories, Oceans can Have Them too." *Ocean Memory*. https://storymaps.arcgis.com/stories/128bbe084ac04b01a3e 02deb350a2219.

Walcott, Derek. 1986. *Derek Walcott: Collected Poem 1948–1984*. Toronto: Faber and Faber.

Weber, Samuel. 2021. *Singularity: Politics and Poetics*. Minneapolis: University of Minnesota Press.

Westphal, Bertrand. 2011. *Geocriticism: Real and Fictional Spaces*. Translated by Robert T. Tally Jr. New York: Palgrave Macmillan.

Whitehead, Alfred North. 1978. *Process and Reality: An Essay in Cosmology*. New York: The Free Press.

Winther, JanGunnar Minhan Dai,Therese Rist, Sandra Whitehouse, Alf Hakon Hoel, Yangfan Li, Ami Trice,Karyn Morrissey, Marie Menez, Leanne Fernandes, Sebastian Unger, Fabio Scarano, and Patrick Halpin. 2020. "Integrated Ocean Management for a Sustainable Ocean Economy." *Nature Ecology and Evolution* 4: 1451–1458. https://doi.org/10.1038/s41559-020-1259-6.

Woodard, Ben. 2013. *On an Ungrounded Earth: Towards a New Geophilosophy*. Brooklyn: Punctum Books.

Zalasiewicz, Jan. 2008. *The Earth After Us: What Legacy Will Humans Leave in the Rocks?* New York: Oxford University Press.

Index

Milton Keynes UK
Ingram Content Group UK Ltd.
UKHW031137141024
449569UK00006B/115